The Sapphire Planet

THE SAPPHIRE
PLANET

Url Lanham

COLUMBIA UNIVERSITY PRESS
NEW YORK • 1978

550
L287

Library of Congress Cataloging in Publication Data

Lanham, Urless Norton, 1918–
The sapphire planet.

Includes index.
1. Earth sciences. 2. Earth. I. Title.
QE31.L33   550   77-13160
ISBN 0-231-03956-5

Columbia University Press
New York and Guildford, Surrey
Copyright © 1978 Columbia University Press

Peace in men,
Good will toward Earth.
            M. L.

Contents

The Sapphire Planet

INTRODUCTION

AFTER AN ACTIVE but carefree life of several billion years, the earth, with the appearance of mankind, has become self conscious, aware of her past, thoughtful of the future. She has seen herself in a mirror, in the photos taken from spacecraft. She is a sparkling gem of a planet with swirling veils of white shining clouds.

The self-image of the earth has grown and changed with the quarter-million-year history of human knowledge. During the immense preliterate era it was the land of the Tribe, whether jungle, desert, or the sounding seashore. With the written word, and travels over seas and continents, a civilized elite fashioned a universal picture. The image of the earth came to be shaped according to the cosmopolitan experiences of science. From the elite, the picture is reflected to the Tribe in books, schools, radios, electronic screens. Although always to a degree subjective, the human consciousness of the nature of the world is now solid with innumerable experiences, with hard contact that shapes both the people and the land.

ONE

·

THE OASIS

THE VISIBLE UNIVERSE is sprinkled with shining, spherical atomic furnaces, the stars. According to a recent educated guess, about one-third of the stars are the centers of flat, rotating discs of cosmic dust and debris, in which are embedded one or more spherical planets. Such a star and its attendant planets make up a solar system. The nuclear star of our solar system is the sun, or Sol. A superficial glance at our solar system would perhaps see it as a sun and two planets, Jupiter and Neptune, which are so much larger than the rest that they contain 20 times the volume of the other seven planets combined. Orbiting about 93 million miles (150 million kilometers) out from the sun is the earth. Five times farther from the sun is the immense planet Jupiter. It is believed that halfway between our sun and the nearest star there are in orbit some billions of tailless comets, which are very cold snowballs, at perhaps $-455°$ F ($-270°$ C, almost absolute zero), and maybe averaging half a mile (a kilometer) in diameter. Occasionally one of these, disturbed by gravitational fields of nearby stars, is sent into the vicinity of the sun, where part of its substance vaporizes into an immense luminous tail pushed outward by the solar wind.

Of all the planets in our solar system, only the earth has a surface temperature (average, 72° F or 22° C) which is both relatively stable and mostly within the range at which water exists as a liquid. Venus, nearer the sun, has a surface temperature of about 900° F (500° C). The small amount of water present is a hot sulfurous steam, and perhaps there are ice crystals in the frozen clouds at the top of the atmosphere. Mars, on the side of the earth away from the sun, has almost no water, and the mean surface temperature is −10° F (−23° C). Liquid water may exist temporarily and locally at sun-warmed places on the surface. On the cloudy surfaces of the large planets beyond Mars the temperatures are even lower—on the most distant planets, not much above the −455° F (−270° C) of outer space.

Only the earth, of all our planets, has large volumes of water at the surface. Convenient for us, and all other living creatures. Water is essential to the existence of the mazy architecture of the intricate molecules whose activity produces life.

Is the earth also the only oasis in the universe? There are billions of stars in a galaxy, and there are galaxies at least by the billions. Theories of resonance in the aggregation of discs of debris around a sun, although still primitive, suggest that it would be by no means unusual to have an earth-sized planet at an earthlike distance from a central sun. However, a here-and-now in our solar system means little in terms of a there-and-now at interstellar and interglactic distances. The barriers of space and time are difficult to penetrate, and the question at the beginning of this paragraph has little meaning for us, or for me at least, at the time of writing.

The Subtle Substance

Water, or liquid dihydrogen oxide, is according to its textbook description colorless in thin layers and blue-green in thick ones. There is a story that the eminent nineteenth-century German scientist Hermann von Helmholtz had concluded from its behavior in electric fields that water was necessarily blue. To demonstrate for his students that water was inherently blue, not just reflecting the azure sky, he is said to have set up a long glass tube filled with distilled water, but when he looked through the apparatus, it did not look blue to him. So for the benefit of all, he added a tinge of blue dye. Now, von Helmholtz is so towering an intellectual figure in both physics and the physiology of sense perception that the story is likely apocryphal; but on the other hand it might be an unexpected glint of humor.

Nevertheless, the chemists are right, and the intense blue of the ocean as seen from a boat far beyond the turbid coastline is the color of the immense volume of water beneath. Since the ocean covers 70 percent of the earth, it should have come as no surprise that ours is a sapphire planet, yet the first photos taken far out in space by the men who coasted to the moon astonished us with the image of the earth as a glowing, gemlike blue planet, banded with gleaming white clouds of water vapor.

Liquid water is a light, mobile, clear substance, flowing in the infinitely varied patterns of brooks, rounding into drops of rain that refract sunlight into rainbows; at subfreezing temperatures it crystallizes into fragile snow-

flakes. It is a remarkable chemical compound. "Strikingly
different from other substances," says the chemist.

Water is one of the few compounds, if not the only
one, that is a liquid, solid, or gas at "ordinary" temperatures, the temperatures at which we usually function. It is
a gas, in varying proportions, in the atmosphere. Ice forms
when the temperature drops below 32° F (0° C), and
crusts the northern and far southern lands in winter. It
forms layers miles deep on Greenland and the Antarctic
continent, holding in cold storage most of the fresh water
of the world. Liquid water exists, at the atmospheric pressure of sea level, only in the narrow temperature band of
32° to 212° F (0° to 100° C). In the solar system, as a
whole, omitting the temperatures in the interior of the
sun, the span of temperature ranges from −455° F (−270°
C) in outer space to nearly 11,000° F (6,000° C) at the
surface of the sun.

The properties of water itself help maintain a narrow
range of temperature on the surface of the earth. Water
has a high specific heat, which means that a relatively
large amount of energy is needed to raise its temperature.
It takes 5 times the amount of energy to raise the temperature of a given weight of water by 2° F (1° C) as to heat
the same weight of granite by that amount, and 1,300
times the energy to heat the same volume of air by the
same amount. In this way, variations in temperature
caused by varying cloud cover, air circulation, and sea-
sonal changes in day length tend to be smoothed out by
the oceans and large lakes. Rising temperatures melt ice,
but when ice melts it absorbs large quantities of heat from
its surroundings, thereby moderating the very temperature change that is producing it. When water freezes, it
gives up heat to the environment.

The delicately poised balance between water, ice, and

vapor makes the surface of the earth a constantly changing scene. The mighty clouds are fragile structures that
change shape at the whim of minute changes in humidity
or temperature of the air. They carry water from the sun-
warmed surface of the sea over the land, where rain is
gathered into rivers that carve the surface into a changing
and varied landscape. Mountain snows pack into glaciers
that pluck away masses of rock to leave craggy peaks.
Along the shores of continents and in river basins, sand-
and-mud-laden water deposits thick layers of sedimentary
rocks.

Water is a sensuous, neutral liquid that affects the
skin only by light touch. In character with the fact that we
cannot taste or smell it, or feel any corrosive effect when it
touches our skin, water is chemically quite stable. This is
because the atoms it consists of lie in a deep energy well.
Water can be synthesized by touching a match to a test-
tube mixture of hydrogen and oxygen. The two combine
instantly with a sharp explosion that yields a large amount
of energy. Droplets of water bead the walls of the tube.
According to the law of conservation of energy, equal energy must be applied to the water to convert it back to
hydrogen and oxygen. Since this amount is large, the energy well of its atoms is deep. Water rarely finds itself in
contact with anything at a high enough energy level that
it will react chemically. There are exceptions, such as sodium metal, which is obtained at great energy cost by the
electrolysis of sodium chloride. One of the hazards of certain kinds of nuclear power plants is that the heat from
the nuclear furnace is carried to the water (which is converted to steam) by molten sodium metal. Should the
channels carrying the sodium break down, the metal and
water would combine with explosive violence.

Although water is soothing, bland, and chemically

very stable, it is at the same time a most subversive and insidious substance. This results from its unusual physicochemical properties. It dissolves more kinds of substances than any other liquid and has been called the "universal solvent." It is likely that every element known on earth exists in solution in seawater.

Because it carries two strong and opposite electric charges, which are widely separated on two sides of the molecule, water has the property of dissolving (or prying atom-sized particles from) such substances as table salt. The "sharp elbows," or as the physical chemist puts it, the large dipole moment, of the constantly jostling water molecules ease apart the sodium and chlorine of the salt. The products of this dissociation are not atoms but ions, which are atoms with fewer or more electrons than they normally have. The electrically charged ions are chemically reactive.

So active a solvent is water that pure water is essentially nonexistent in nature. Even rain drops contain dissolved carbon dioxide and other substances from the atmosphere when they hit the ground. When water becomes a solution of chemically active ions, it becomes extremely corrosive. Water carrying carbon dioxide dissolves out caves in thick layers of limestone rock. Water in deep, hot layers of rock far below the surface is a devil's brew of dissolved minerals that are considered insoluble at ordinary temperatures and pressures. In cracks and cavities, as they work their way upward, these corrosive waters deposit veins of metal-rich ores and the elegant clean-edged crystals prized by collectors.

Living things, however, find water to be benign rather than a corrosive enemy, and are in fact themselves mainly water. Many organic compounds are completely unaffected by water, such as various waxes or the oils on the surface of our skin. Water rolls off the duck's back because preening has spread oils on the feathers. Flocks of unwanted starlings have been killed by spraying them with a detergent that breaks up the oily substances so that the birds die in wet cold weather of hypothermia, deprived of their insulating, waterproof blanket.

Other organic compounds, often molecules of great size and complexity, influence the surrounding water molecules (which generally make up about 70 percent of the total mass of living substance) in ways that serve the purposes of life. Water in the vicinity of such giant molecules is not an amorphous mass of particles in random motion, but is bound into a variety of permanent and semipermanent configurations that are a part of the complex dynamic architecture of the living cell.

Unlike the high-energy-requiring situation in the laboratory, water is split into its components by the living world in subtle fashion, using the gentle energy of sunlight. This takes place in photosynthesis. The energy that is potential in the existence of separated hydrogen is released by the biological process called respiration. Here the oxygen of the atmosphere combines with hydrogen of the organic compounds, hydrogen that was derived originally from water by photosynthesis. This organic combustion takes place step by step, each yielding only a fraction of the total energy finally obtained. The recombination of hydrogen and oxygen in respiration of course produces water.

Figure of the Earth

For nearly all the thousands of generations of the human race that have lived and died, the earth was a great mother of unknown dimensions and form. Or,

rather, of protean form, since she was described in countless ways in the legends of primitive peoples. The modern
conception of the earth's form became a permanent fix-
ture in the world's stock of ideas by the sixth century B.C.,
when the Greek mystic and scientist Pythagoras paid special attention to what he said was an old Egyptian notion
that the earth was a sphere, suspended in space. His students added the idea that it turned on its axis each day,
making the sun and stars rise in the east and set in the
west.

Civilization was by this time several thousand years
old, and there was already a long tradition of correlating
seasons with the behavior of the sun and moon and stars.
This was early expressed in mathematical form, both "on
paper" and in primitive analogue computers consisting of
sighting points set in the ground in certain geometric patterns. These gave records of the position of heavenly bodies on the horizon at different times of the year. Refine-
ments in calendars gave each day of the year a name.
Shadows cast by vertical markers during the daylight
hours quantified the time of day, and their length at noon
the season. Scholars in distant places could compare
notes. In this way Eratosthenes, of the third century B.C.,
was able to solve the problem of the size of the earth.

Eratosthenes knew that in the town of Syene (the
modern Aswan), some 500 "stades" directly south of Alexandria, on a certain day the sun at noon was exactly over-
head, so that a vertical pole cast no shadow. At Alexandria, on the same day, the noon shadow of a vertical
pole had a certain length indicating, by way of elementary
geometry, that the circumference of the earth was about
50 times the distance between Syene and Alexandria.
This would come out to 25,000 miles (40,000 km), a good
estimate.

During the Dark Ages the concept of a spherical

earth was lost to much of the European community, but it was revived in the Renaissance and made the subject of refined observations during the intense worldwide geographic explorations of this time.

In the 1670s the French astronomer Richer, sent to South America to make some astronomical observations, observed that his pendulum clock lost about two minutes a day unless he shortened the pendulum. The rate of swing depends on the force of gravity, so the observation meant that gravity was weaker at the equator than at high latitudes. Isaac Newton thought the effect could result either from the centrifugal force of the earth's rotation, which would make an object weigh less, or from a bulge in the earth at low latitudes, which would put the pendulum bob farther from the center of the earth and therefore make it more weakly attracted. It took three more centuries of observation to demonstrate that both factors produce the effect, in combination. The "figure of the earth," as it is called, shows an equatorial bulge that makes the circumference around the equator 40 miles greater than the circumference measured through the poles.

During recent decades it has been necessary to find the dimensions of the earth, and distances between far distant points, to within a fraction of a mile. This information makes it possible to send intercontinental missiles to their destination, and to learn more astronomy. These refined measurements result from methods of determining the time of flight of electromagnetic waves, usually light or radio waves (which travel in a vacuum at 186,171 miles, or 299,792.8 km, per second), between sending, reflecting, and receiving points. Such measurements of the distance of artificial satellites orbiting the earth give very accurate outlines of its shape.

It was not possible to determine the mass of the earth until the formula for the force of gravity, which holds the solar system together, was discovered. When we fall, or lift a heavy stone, we think of gravity as a very powerful force indeed, but this is only because the earth is very massive compared to us weak and fragile creatures. The gravitational attraction between any objects we can conveniently manage in the laboratory is extremely weak.

But it is not quite imperceptible, as the British physicist and chemist Henry Cavendish (1731–1810) was able to show in a series of arduous experiments whose results he published in 1798. Using a complicated contraption consisting of heavy lead weights suspended on thin wires, shielded from air currents and changes in temperature, and observed from a telescope outside the room, he was able to see the force of gravity in action. In essence, he observed the amount by which a sphere of lead weighing 350 pounds attracted a small sphere of lead 2 inches (5 cm) in diameter, weighing about 1¾ pounds (800 g). By a complicated sequence of mathematical reasoning, his experiment actually gave the density of the earth, which came out to be 5.48 (that is, the earth is five and a half times as dense as water). The total mass of the earth came to something like 6 billion million million metric tons.

Since the average density of rocks that we can observe at the surface is only 2.8 g/cc (grams per cubic centimeter), the center of the earth must be very dense. The earth has the greatest average density of any of the planets, with Mercury and Venus a close second and third. Gigantic Jupiter has a density of only 1.33 g/cc, not much more than that of water.

A Place in the Sun

For centuries before the time of the scientific Greeks, the intricacy and precision of the movement of sun, stars, and planets through the heavens gave pleasurable employment to the scholars in the ancient civilizations of the eastern Mediterranean. These early astronomers had been content to take the movements without question and to use them in timing the annual rituals of agriculture, in divination, and in religious rites. They made little or no effort to explain these phenomena of the heavens in terms of earthly physical processes.

The Greeks, probably because of a more loosely organized state apparatus that did not depend on a rigid theological bureaucracy, differed from their neighbors in taking a matter-of-fact attitude toward the objects in the skies. Some of their scientists tried to make three-dimensional models, with tangible spheres or points of light that moved according to definite mathematical rules, to explain the universe.

The Greek astronomer Aristarchus of Samos, who lived in the third century B.C., developed an essentially correct method for determining the size and distance of the sun and moon. However, it was difficult to make the accurate measurements of movement that were required. His estimate for the distance of the moon, 300,000 miles, was good, but he thought the sun much smaller and closer than it is. He believed that the earth and the planets revolved about the sun, and thus his model, with a mobile earth and great distances, was a modern one.

However, Aristarchus turned out to be nearly 2,000 years ahead of his time. The most accomplished of the Greek astronomers, Hipparchus (who flourished toward the end of the second century B.C.), accumulated prodigious quantities of relatively accurate data on the movements of stars and planets. He opted for the more popular and psychologically more reassuring theory that the earth stood fixed at the very center of the universe, with the heavenly bodies revolving about it. In fact, this model is still found to be convenient in textbooks on navigation, in which a "stellar sphere" is considered to rotate daily about a fixed earth (it is when the sun and planets are taken into account that the Hipparchus model becomes unmanageable). Hipparchus made a lasting impression on astronomy, and his geocentric theory was held even by the educated Moslems until the sixteenth century of our era.

Archimedes, of the third century B.C., the best of the Greek engineers and mathematicians, had written an account of Aristarchus' sun-centered solar system. This disappeared from the European scene, along with most other Greek scientific writings, with the advent of the Roman Empire.

The scientific renaissance of the sixteenth century was accompanied by the recovery of many of the classical Greek scientific writings. They had been preserved throughout the European Dark Ages in Arabic translations on the south side of the Mediterranean. These were now being translated into Latin. When the works of Archimedes were brought to light, his account of the sun-centered system of planets was widely discussed in Europe and, of particular importance, attracted the attention of the Polish astronomer Nicolaus Copernicus (1473–1543) and of Galileo Galilei (1564–1642), an Italian.

Copernicus, a busy and competent professional cleric in the Catholic Church, was only by avocation an astronomer. Well trained in his youth in both astronomy and mathematics, he spent much of his life rearranging the data of astronomy in such a way as to support the classic heliocentric theory of Archimedes. The work that summed up his studies, *Concerning the Revolutions of the Heavenly Bodies,* was published in 1543.

Galileo was not primarily an astronomer, but his pioneering use of the telescope to examine the heavens proved to be more effective than the intellectual work of Copernicus in jolting people's minds out of accustomed ways of thinking. Galileo could see four small moons circling Jupiter, just in the way Copernicus had said the planets orbited the gigantic sun. As the Copernican theory predicted, he could see that Venus, moving between the earth and the sun, had phases like the moon. As seen by Galileo through his telescope, the hazy Milky Way turned out to be made of countless bright stars, giving a stomach-sinking glimpse of infinity. It is said that some clerics refused to look through Galileo's telescope, and others looked but protested that the supposed objects were produced by the instrument itself.

Galileo early became convinced that Copernicus was right, and supported the Copernican theory in a work published in 1613 entitled *Letters on the Solar Spots.* Galileo was reprimanded by the Pope but in 1630 threw down the gauntlet with the publication of his chief work, *Dialogue on the Two Chief Systems of the World.* The Roman Church, which had taken Copernicus rather coolly, now declared Galileo, his theories, his lectures, and his books (and perhaps his telescope) to be illegal, and he narrowly escaped death.

The most competent astronomers of Copernicus' time could not accept his ideas because the more accurate observations on planetary movements made possible with improved instruments did not match his theory that the earth and other planets moved in circular orbits about the sun. Later, however, Johannes Kepler (1571–1630) found that if the planets were considered to move in ellipses rather than circles, then the sun-centered theory of Copernicus fitted the new data quite well. The change from circle to ellipse is a trifling one in a mathematical sense. Both are sections through a cone, with similar mathematical properties that made possible the rather simple explanations developed later for the dynamics of the solar system. In the earth-centered theory of medieval times, a theory held even by the sophisticated Moslem scientists, the planets took extremely complicated paths, moving in circles adorned with minor reverse loops that make little sense dynamically.

Galileo's specialty was physics, and he was concerned with one of the important problems of war-making nations: how do you calculate the path of ballistic missiles, how do you hit the target with an artillery shell? By use of many experiments with heavy spheres in motion, in which friction from air resistance and other sources was minimized and its effects discounted, he was able to propound various laws of motion. One of these was that an object once set in motion at a given velocity tended to stay in motion at that velocity in a straight line, unless acted upon by an outside force. Another was that an object fell toward the earth in a straight line at a constantly increasing velocity, the acceleration being about 32 feet (9.8 m) per second per second. This force of gravity is constantly operating on the free-falling object, hence the acceleration.

Galileo reasoned that a cannonball in flight was subjected to two forces: First, the initial force applied by the explosion of the gunpowder which, if no other forces acted on the cannonball, would cause it to travel in a straight line out to infinity. The second and only other force was the tendency to fall toward the earth in a straight line, which pulled the cannonball toward the earth at an accelerated rate. The resultant path, Galileo was able to show, was a parabola which, like the circle and the ellipse, is a conic section although open rather than closed.

These conceptions were applied to the question of what caused the planets to follow their observed orbits, and the problem was solved later in the seventeenth century. Probably one of the first working models of a planetary orbit was supplied by that versatile man of many sciences, Robert Hooke (1635–1703) of England. Using a pendulum consisting of a weight suspended on a wire, he started it moving with a sideways push. The weight then moved, slowed down only by air resistance, in an orbit that was circular or elliptical (nearly). Actually two forces and inertia were involved: the pull of the wire toward the support, the gravitational pull of the earth, and the initial need for a push. Hooke simplified the problem by ignoring the earth's effect. He compared the moving weight with the planet, and the wire (with its continually exerted force) with the gravitational pull of the sun.

By this reasoning Hooke was extending to the heavens the same sort of forces that could be observed on earth, although what set the planets moving remained a mystery. Being no mathematician, Hooke was unable to quantify his theory, a formidable project that was carried out by his contemporary, Isaac Newton (1642–1727), one of the inventors of the calculus. It was clear that if gravity was the force that held planets in their orbits, it had to apply to all objects (including apples, so the story goes), with more massive objects exerting greater force than lighter ones. Also, gravity had to become weaker with increasing distance. Using the hypothesis that the force of gravity decreased as the square of the distance, Newton tried it out on the orbit of the moon around the earth. Knowing the distance of the moon, he calculated the rate of acceleration of fall toward the earth if the rate known at the earth's surface decreased as the square of the distance of the moon, a quarter of a million miles away. He found that the moon's departure from a straight line in a given time (that is, its fall toward the earth away from its theoretical straight-line flight away from the earth into outer space) corresponded to the amount he had calculated. He also estimated that if the moon circled the earth just above the highest peaks it would take about an hour and a half for a complete orbit, a prediction borne out by Sputnik.

Newton depicted the solar system as a gigantic machine in which the planets eternally orbited the sun, a machine that operated with no guidance other than the force of gravity. The earth and other planets could be regarded as Galilean cannonballs, set in motion at the moment of creation and falling forever toward the sun in closed conic sections.

In the eighteenth century the astronomer Pierre Simon de Laplace (1749–1827) considered the question of whether or not the solar system could keep operating indefinitely. Refined measurements showed irregularities in the movement of planets not explained by Newton. Would they cause the machine to break down? Laplace's analysis showed that in fact these irregularities, like all other known movements of the planets, repeated themselves, even though some cycles were calculated to be 900 years long. The machine did seem to be flawless and eternal.

Did it, and does it, really make any difference to most of us to know that the earth moves, that it turns on its axis once a day, and circles the sun once a year? Sherlock Holmes scornfully remarked that such knowledge was useless to himself; he cared not a particle for astronomy. The painter Vincent van Gogh thought the phases of the moon were caused by the shadow of the earth. A modern graduate student in biology who was doing a thesis on aquatic invertebrates had, at least at the time of his comprehensive examination, not the slightest conception of the structure of the solar system or its operation.

At the time of these fundamental discoveries, it did make a good deal of practical difference whether or not the earth stood still. Those who said it did not were, in some countries and at some times, believed to represent a danger to public order. The Roman Church had in many instances staked its credibility upon these astronomical matters, and the fact that it proved itself highly fallible contributed to the political instability of the times, since the Church was an important political power.

A Place in the Universe

So far as factual evidence went, astronomers just after the time of Newton could suppose the universe to consist of a solar system some 2 billion miles (3.2 billion km) in diameter (Saturn was then the outermost planet known) encased in a thin shell dotted with luminous points called stars. The reason for thinking that the sphere of stars had no depth was the fact that even when the earth was on opposite sides of its orbit (nearly 200 million miles, 320 million km, in diameter) the stars remained in the same relative positions. Were some farther away than others, their relative positions should change because of parallax, just as a distant tree seems to change its position relative to an even more distant hill as we drive by in an automobile.

It could, of course, be stated as a hypothesis that the

stars were scattered through a very large or even infinite universe, with the nearest stars so far away that the best instruments could not detect parallax. The idea that the stars were distant suns, scattered through space, was an old one. The Italian philosopher Gordiano Bruno (1548–1600), well-versed in the ancient pagan wisdom, was burnt at the stake by the Inquisition for holding this and certain other opinions.

The picture of the universe as a solar system in a star-studded bubble began to dissolve in the nineteenth century, when comparison of new and old star maps showed that some of the brighter stars were moving relative to one another, for reasons other than parallax. The fainter stars appeared to remain motionless. This was taken to mean that the faint ones were too distant to detect movement; the brighter, moving stars were nearer the earth. Soon afterward, instruments were devised that were good enough to detect true parallax of the nearest stars. The universe beyond the solar system was sprinkled with stars in all directions. In 1838 it was calculated that the nearest star was about 26 million million miles (42 million million km) away.

A determined, pioneer explorer of this new universe was William Herschel (1738–1822), a German musician who early in his career came to England. He had his musical training from his father. He taught himself mathematics, and was entranced by the science of astronomy. While still a successful musician, he began to find the means to carry out his dream of exploring the universe.

In his day the lumps of glass of the kind needed for aberration-free lenses could only be made 3 or 4 inches (7–10 cm) in diameter. Lenses of this size could not gather enough light to clarify images of objects that might lie in the farther reaches of the universe. Newton had shown that the way to get around the difficulty was to use a concave mirror rather than a refracting lens for astronomical telescopes, but no one had been able to make a mirror appreciably better than the available lenses.

Herschel took up the problem of making better telescopic mirrors, bringing his 23-year-old sister Caroline from Germany to help (she later became a well-known astronomer in her own right). While still making a living as a musician, he made hundreds of mirrors, some of them giving a better view of the stars than anything made before. To hollow out and polish a mirror from a disc of metal meant that Herschel had to walk for long hours around the mirror, holding against it a cloth with abrasive. His sister fed him as he walked. The largest mirror that he ground had a diameter of 4 feet (122 cm) and a focal length of 40 feet (12.2 m), but some of his smaller mirrors were of better quality and had better resolution.

Herschel, while still an amateur (an amateur astronomer has been defined as one who is genuinely sorry when the sky is cloudy), at 43 years of age discovered the planet Uranus, thereby doubling the known size of the solar system. The following year he lost his amateur status when he was appointed court astronomer by the King of England.

Without a telescope, one sees the stars as rather evenly distributed through the sky, but with an astronomical telescope the picture is much different. The Milky Way, a luminous band arching across the heavens, little known to city dwellers because it is obscured by reflection of city lights in the night sky, becomes a multitude of stars. As a result, most of the stars seem to be crowded into one plane, somewhat as the planets of the solar system move within a relatively flat band. Herschel decided that the Milky Way represented all of the universe, and that the sun was merely one of the stars in this vast disc-shaped constellation, which came to be called the Galaxy. He estimated that the stars were some trillions of miles apart and that the number of stars in the Galaxy, and in the universe, could be approximated at 300 million. Later in his career, he came to believe that he had probably not plumbed the depths of the universe and that there might be other galaxies of stars in existence.

Herschel's telescopes showed that the universe contained more than stars and planets. He could now see in considerable detail some large, fuzzy objects which had been called nebulae (Latin for "clouds"). He increased the number of known nebulae from just over a hundred to over 2,500. Nebulae are a varied and beautiful lot. Some are bright, diffuse clouds, with streamers that are doubtless in graceful movement over a time scale of millions of years. Others are rings, sometimes with bright eyelike centers. One, Andromeda, is a disc seen nearly on edge but in which a pair of spiral arms can be made out.

So dusty is the Milky Way that Herschel could not see enough stars to determine the detailed structure of the Galaxy, but could only say that it appeared to be disc-shaped. In the twentieth century, telescopes that could focus radio waves were invented, and since these waves travel unhindered through interstellar dust clouds, and since the neutral interstellar hydrogen gas sends out radio waves, the mapping could be done more accurately. Our galaxy has been found to consist of three spiral arms lying in a single plane, which emerge from a bulging center. The sun is a star rather far out on the second turn of one of these arms.

Various technical and theoretical developments have made it possible to estimate distances of stars in at least a tentative way, by analyzing their brightness and the spectrum of their light. In this way, a large number of very distant objects were discovered that were thought to be so far away that no imaginable single star could give enough light to be seen. These must instead be galaxies, each made up of billions of stars. Some of the closer ones, like Andromeda, show enough detail to support this hypothesis, and Andromeda itself resembles our own galaxy in form. It is now believed that there are 800 million galaxies within reach of the 200-inch telescope at Palomar—Herschel would be astounded.

The basic units of the universe appear to be the galaxies, perhaps scattered more or less at random, or perhaps arranged in groups.

The universe seems to be a restless place, although size is on a scale so majestic that motion is slow and ponderous. Close at hand the moon and planets move deliberately through space. Only the flashing streak of an occasional meteor gives a hint of the speed and power residing in space.

It is a strange observation that all but the nearest galaxies seem to be moving away from our own. The farther the galaxy, the greater its speed. It is not that the Milky Way is a center for this universal expansion, but that the universe seems to act as an expanding sphere. No matter which galaxy was picked as the observation point, the more distant galaxies would appear to recede more rapidly. This expansion is deduced from the fact that light from distant galaxies is shifted toward the red end of the spectrum, where the wavelengths are longest and the vibration frequencies lowest. The same phenomenon is seen, or rather heard, in sound waves. A train whistle receding down the tracks drops in pitch as it goes by, because the wave lengths shift into lower frequencies and stretch out into longer waves, in the perception of the observer. This is called the Doppler effect. The Doppler effect also applies to electromagnetic waves, including light waves, as can be demonstrated in the laboratory.

There is some controversy about whether all the red shift observed in light from distant galaxies can be attributed to the Doppler effect. It may be that other factors operate over the immense distances of space. According to the Stokes law, light absorbed by atoms is usually reemitted at a lower frequency. Space is filled with a sparse population of atoms. There may be unknown electromagnetic forces that are effective at these distances. We are seeing light that is billions of years old when it reaches the earth from a distant galaxy. However, the consensus is that the effect is a true Doppler effect, and that the galaxies are receding at a rate that is roughly proportional to their distance.

A Universe, of Sorts

Although awe-inspiring because of its vast size, the universe seen through telescopes was straightforward enough. It consisted of material bodies, the planets, orbiting about stars grouped into galaxies. Not all of the visible structures were reliable solid objects, but at the worst they seemed to be clouds of gas of a kind more or less understood in chemistry laboratories. What was to make the universe again a strange and unsettling place was not the study of the heavenly bodies but certain laboratory studies concerning the effects of subatomic particles in motion.

Probably most of the trouble began when methods were developed for taking apart atoms in controlled fashion and working with one of these components in pure form and in large quantities. These easily obtained components were electrons, which circle the nuclei of atoms at relatively vast distances, and are easily detached from the nuclei. Early in the nineteenth century it became possible, mainly by chemical means, to pump quantities of electrons through copper wires. The result is an electric current.

An English chemist and physicist, Michael Faraday (1791–1867) discovered that the empty space around a copper wire is altered when an electric current begins to flow through it. A test wire in the vicinity, not connected with the original wire, will also show a brief surge of electricity. If the original current is converted into alternating current, so that its direction is reversed (it can be reversed thousands of times per second), the test or receiving wire also shows an alternating current. In modern terminology, it is immersed in a field of electromagnetic waves, or radio waves.

Obviously, this strange behavior of space around moving electrons offers a mode of communication. Today electric currents are manipulated in fantastically complex and subtle ways at transmitters. The resulting force fields can cause radios and television receivers to produce sounds and moving images.

Actually, communication by electromagnetic waves has existed for as long as living creatures have had eyes. The vibratory force field consists of light waves, and their source is the electrons that orbit the nuclei of atoms.

The empty space of the universe is a dynamic and complex configuration of electromagnetic waves, including X-rays, light and radio waves, and possibly gravity waves. Rather than being inert, empty space is the theater of powerful forces, or perhaps should be said to consist of force.

During the nineteenth century physicists determined the speed of light, which was already known to be so high as to be instantaneous for practical earthly transactions. An early successful method was to measure the displacement of a light beam reflected from a rapidly spinning mirror. The speed of light through the atmosphere is 186,282.4 miles (299,792.8 km) per second. The speed of radio waves, determined by an entirely different method, is the same. Transparent substances that are very much denser than the atmosphere, such as liquids or solids, slow up light by a small amount because of interaction with atoms along the way.

It is impossible for any material object or any kind of signal to move faster than the speed of light; that is, it represents absolute velocity. If a measuring instrument is moved toward a stationary light source at any velocity whatsoever, it still measures the velocity of light at the same value as if the instrument were also stationary. Early in the twentieth century this result was predicted on theoretical grounds, and it has also been shown in laboratory experiments. It makes quite impossible the task of choosing any one point in the universe as being at rest.

The invariance of the speed of light leads to several surprising consequences, which were developed into Einstein's special theory of relativity, published in 1905. Most of these have since been proved experimentally. Among them is the fact that with increasing speed the mass of an object increases. As measured by clocks, orbital frequencies of electrons orbiting atomic nuclei, and biological processes (or, as Bertrand Russell suggested, how long a cigar lasts), time slows down as perceived by a stationary observer. For example, at about 90 percent of the speed of light, to the bystander a clock would appear to take 2 seconds to tick off 1 second, while at the full speed of light the clock would seem to stand still. To an observer on the 90 percent ride, it would seem that the clock of the stationary observer was slow. Also, objects become foreshortened in the direction of motion. If one earthlike planet flashed past another at the 90 percent speed, an observer on either would see the other planet as an ellipsoid only 12,000 miles (19,300 km) in its short diameter, and its mass would be twice as great as that of the planet under the observer's feet.

This peculiar universe is rendered even more alien by our modern mathematical description of subatomic particles. According to theoretical work done by the physicist Werner Heisenberg (1901–) in the first third of the twentieth century, it is impossible to describe an electron as a particle, or as some had tried to do, as a wave. A graduate student in physics once told me that you knew you had "arrived" when you understood that it makes no difference whether the electron is a wave or a particle; that you merely used the concept most useful in solving the problem at hand. Further, Heisenberg's "uncertainty principle" stated that it was inherently impossible to determine both the position and motion of a subatomic particle: in determining one, you destroyed the possibility of determining the other. Thus, at this size level nature is in a fundamental way unpredictable.

At the prosaic technical level, there appear to be two kinds of approaches that could be used to penetrate the barrier of the finite speed of light and of all other known forms of movement or transmission. First, at very high speeds, approaching the speed of light, it is theoretically possible to reach a star or planet thousands of light-years distant during the lifetime of an individual, according to the usual interpretation of the effect of relativistic speeds on time. This would of course mean nothing to those of us left on earth, since hundreds of generations might have lived and died before a signal of the traveler's safe arrival reached the earth. Secondly, the science of cryogenics or cryobiology might have developed sufficiently that life could be slowed down to the point where long-term space travel would be possible.

At the level of science fiction (or perhaps it should be called "non-science" fiction), the ordinary restraints of science are abandoned and extrasensory perception, teleportation, and a multitude of other ways of crashing or stealing through the barriers of space and time have been imagined. The universe is a peculiar place, and we may well see some unexpected developments, but they probably will not correspond at all to our wishful thinking.

Writers of science fiction, like theologians, generally find it impossible to reconcile themselves to the idea that man is on his own. At least this is the image that many of them project in their writing. Arthur Clarke's *2001* has extraterrestrial beings coming to earth to program apes to convert them to human beings. In *The World of Null-A* by A. E. van Vogt, the heroine says, "I don't know how man got to Earth. The monkey theory seems plausible only when you don't examine it too closely."

The extraordinarily strong drive to believe that there is someone in charge more intelligent than ourselves is not very reassuring, seen as a kind of Gallup poll on human competence.

TWO

·

INTERNAL PHYSIOLOGY OF THE EARTH

The Warm Body

IT HAS LONG BEEN KNOWN that the temperature increases as we dig deeper into the earth. This becomes a practical problem in deep mines. The Comstock silver mines of Nevada reached a depth of over 3,300 feet (1,000 m). Miners here worked with rock as hot as 160° F (70° C), and many died from the heat. South African gold mines in the Witwatersrand are down to 10,000 feet (3,050 m). Only cooling systems make it possible to work there. The rate of heat increase varies from place to place, owing to local geological conditions. The more usual gradients average about 1° F per 30 feet (1° C per 50 m). Such measurements apply only to an upper thin layer of the earth's crust. The deepest bore holes have penetrated only

4.3 miles (7 km). It is impossible for temperature to continue increasing at 1° F per 30 feet down to the center of the earth, since the temperature at the core would be implausibly high, such as to destroy the earth by an explosion.

Boiling water (212° F, 100° C) comes out of the ground at hot springs, and glowing molten rock emerges from volcanoes. Direct measurements of flowing lava at the surface give temperatures of about 2,200° F (1,200° C). In the laboratory, solid lavas can be remelted at about 1,800° F (1,000° C). The depth from which the molten lava arises is not known, but probably it does not much exceed 18 miles (30 km). Some measurements, made by indirect means, of electrical conductivity at the boundary between the core and mantle of the earth (at approximately 1,800 miles, or 3,000 km, down) lead to an estimate of 7,000–9,000° F (4,000–5,000° C) at that depth.

As the cold of a winter night or of the winter polar wastelands makes clear, this internal heat is not enough to make the surface of the planet a comfortable place to live. It is the sun that warms us. Heat absorption from sunlight and heat flow from the interior of the earth can be quantified as follows. The flow of heat from the interior through the surface and into the air, both on land and at the sea floor, is slightly over a millionth of a calorie per second per square centimeter, or 6 millionths per square inch. This would supply enough heat to melt a film of ice ¼ inch thick (6–7 mm) each year. Violent geothermal events such as the eruption of volcanoes, if imagined to have their effect spread evenly over the earth's surface, yield only about 1/50 of this amount. The sun, by contrast, provides 3,000 times as much heat to the earth as that from the interior. The result is an average temperature of 46° F (8° C) over the surface of the earth.

The heat from the sun affects only the outer skin of the earth. Some of the energy of sunlight is immediately reflected from the upper atmosphere. The rest is absorbed, providing the energy for the complicated affairs of the surface of the planet—winds, waterfalls, and life. These various forms of mechanical energy are eventually converted into heat, which finally radiates back out to space, to be lost in the intergalactic wilderness.

Little is known about the origin of the earth's heat at very great depths. However, quite a bit is known about the warmth generated in the crust. It is strange to think, during the current debate on the development of nuclear energy to supplement coal and oil, that the furnace heating the ground under our feet is a gigantic nuclear reactor. The part of the earth's crust that forms continents is mainly granite. Granite contains minute amounts of radioactive elements which by nuclear disintegration produce heat energy. The elements responsible for this natural radioactivity are chiefly isotopes of thorium, uranium, and potassium. Their natural rate of decay is such that a cubic centimeter ($1/16$ cubic inch) of average granite yields about 1/100,000 of a calorie of energy per year. So vast is the volume of granite, however, that nearly half the heat flow is produced by this radioactivity.

The lower section of the crust, which is composed of basalt rather than granite, and which forms much of the ocean floor, must also be radioactive but to a somewhat lesser degree. Apparently the greater volcanic activity on the ocean floor compensates to bring the heat flow through the surface to an amount comparable to that on the continents.

Living systems have evolved in the presence of the damaging radiation produced by the radioactive uranium, thorium, and potassium of the earth's crust. This radiation, together with that coming from the sun and the cosmic rays of outer space, is responsible for many of the genetic mutations that occur in plants and animals. Most mutations are probably harmful, although living cells are equipped with repair mechanisms that cope with much of the damage. Other mutations are beneficial because they contribute to the store of genetic variability that is necessary for evolutionary change.

If the radioactive elements could be extracted from, say, a ton of granite, and assembled into a properly designed nuclear reactor, they could yield in a short time the amount of energy in 15 tons of coal. The difference to living systems is that the radioactive process which in nature takes hundreds of millions of years is here compressed into weeks or months. Lethal radiation and radioactive elements pour out in an overwhelming flood that must be kept away from people. Satisfactory methods for achieving this still have not been worked out. The present methods of disposing of waste from nuclear reactors are only makeshift and cannot be continued into an era of large-scale production of nuclear energy. The nuclear plants themselves become overwhelmingly radioactive in a matter of decades and must be fenced off as death zones for centuries or millennia.

An inquiry into the ardent properties of the earth led, by the kind of inevitability one often finds in puns, to the answer of the question, how old is she? Already in the eighteenth century the French scientist George Buffon (1707–1788), who wrote a popular 44-volume encyclopedia of natural history, had decided that the earth had originated as an incandescent sphere. Measuring how fast hot spheres of iron cooled off, he decided that the earth took 30,000 years to become cool enough to support life. In another 30,000 years, animals appeared, some

15,000 years B.P. (before the present). He thought that the earth had cooled off too much to melt lava and that the heat of volcanoes came from burning coal beds.

By the last half of the nineteenth century, many physicists considered the main problems of the physical world to be solved. One of the most confident was the renowned and honored English scientist William Thomson, later Lord Kelvin (1824–1907). Also a very religious man who thought the biological world to be in the capable hands of God, he was pleased to be able to show, as an acknowledged expert on thermodynamics, that the earth could be no older than some tens of millions of years, too young for the kind of slow evolution by natural selection that Darwin saw as the source of living creation. This age was calculated from what was then known about the temperature gradient near the earth's surface. The later discovery of radioactivity of course confounded all simple interpretations of heat transfer within the earth. But remarkably enough it was the radioactive elements that provided a variety of clocks that could be used to date events in earth's history.

There are about 92 kinds of natural elements but many of these come in varieties called isotopes, which are alike chemically but differ in nuclear structure, producing slight differences in mass. The differences in the way the nucleus is put together (the main components are protons and neutrons) also affect nuclear stability. The isotopes that are measurably unstable are called radioactive. In the nucleus, where the particles are packed close together, any kind of rearrangement releases vast amounts of energy. There is some sort of law governing the universe that requires that the greater the proximity of small particles, the more energy is involved. The comparatively feeble energies of chemical reactions involve only the electrons, which are at (relatively speaking) huge distances from the nucleus.

The breakup of radioactive nuclei releases high-powered radiation or fast-moving particles that can be detected by suitable instruments. Disintegration is random and spontaneous. It is not affected by heat, cold, or other usual environmental changes (changes in an electromagnetic field slightly alter the rates of decay in those few instances where radioactivity depends on the capture of electrons by the nucleus). Chemically active substances have no effect on these waterproof and shockproof nuclear clocks.

At one extreme, there are radioisotopes with a half-life (the time required for half the atoms in a sample to decay) of only a fraction of a second; at the other are those such as uranium with a half-life of billions of years. Experts generally agree that the clock can be used to measure periods up to about ten times its half-life. There is in theory a radioisotope available for almost any time span, but in practice some are too difficult to use.

Within a decade after the discovery of radioactivity (late 1890s), the American investigator Bertram Boltwood (1870–1927) used uranium to determine that certain rocks were on the order of billions of years old. It was not until the middle of the twentieth century, however, that scientists developed techniques to use radioisotopes in a large-scale and consistent fashion to establish an earth chronology.

The radioisotope carbon 14 was one of the first to be used. Carbon 14 is produced when cosmic rays (high-speed nuclei of many kinds of atoms) crash into molecules of the upper atmosphere. Neutrons are produced and, when absorbed by nitrogen nuclei, they transform that element into carbon 14. This combines with oxygen to form carbon dioxide. Only an extremely minute fraction of the carbon dioxide in the atmosphere has the radioisotope; nearly all contains ordinary nonradioactive carbon. The concentration of carbon 14 and its decay rate are such that a gram ($1/28$ ounce) of carbon (containing a very large number of atoms—roughly 5 followed by 21 zeros) taken from the atmosphere produces on the average 16 atomic disintegrations per second. At least it used to decay at this rate. Since the advent of thermonuclear explosions, the amount of carbon 14 has almost doubled, raising the decay rate to 28 per second. Standard laboratory instruments are used to detect and record the disintegrations. Determining the radioisotope content of a carbon sample is an exacting process, and the combined laboratories of the world can analyze only about a thousand samples a year.

Carbon dioxide is removed from the air (and therefore from the possibility of accumulating more radioisotope) when wood is produced by living trees. Carbon 14 is thus especially useful in dating archaeological sites, where charcoal from campfires or wood used for construction may be found. The radioisotope has a half-life of about 5,700 years, so that a piece of wood 10,000 years old will have a disintegration rate considerably less than the 16 per second of the carbon dioxide of the atmosphere. By the time the wood is 60,000 years old, the decay rate is too small to give a reliable estimate of its age.

The radioactive clock useful for dating events early in earth's history is uranium, which has two important radioactive forms, uranium 235 and uranium 238. These have half lives of hundreds and billions of years respectively. They decay, by way of intermediates, into two isotopes of lead. Uranium is widespread in rocks of all kinds but usually in small amounts, so that to use these clocks exceedingly sensitive methods for distinguishing between individual nuclei of uranium and lead, and between their isotopes, are needed. The basic instrument is the mass spectograph, in which the atomic nuclei are shot at high velocity through a magnetic field that deflects them in differing amounts, according to their mass.

The uranium–lead clocks suffer from the disadvantage that both elements are leached out or concentrated by ground water. Judicious choice of uranium-containing minerals, and the fact that there are two clocks running simultaneously but at different rates, make it possible to reduce the error.

In recent years another method based on uranium has been developed that gives consistent results. In addition to the usual decay to lead, uranium nuclei of both isotopes disintegrate by fission, of the kind used in atomic explosions. In nature this event is even more rare than the decay to lead, but it occurs often enough that permanent records of fission can be seen in crystals containing only traces of uranium. The nuclear particles that result from fission fly apart with enormous energies and make paths of destruction through the crystal that can be seen with an ordinary microscope if the crystal is first cut and polished, then etched with a solvent. When counted, these fission tracks give a clue to how long ago the crystal was formed. It is obvious that the calculation cannot be made unless the number of uranium atoms in the crystal is also determined. This is accomplished by showering the crystal with neutrons, after the first track count is made. The neutrons produce fission in a known percentage of uranium atoms, yielding a new crop of tracks. The relationship between the number of old and new tracks makes it possible to calculate the age of the crystal.

Although interpreting radioisotope clocks is a complex subject, requiring sophisticated mathematical and laboratory techniques, the variety of methods used provide cross-checks allowing increasingly reliable estimates of age, ranging from the thousands of years of earth's history in which prehistoric peoples have lived back to the lifeless era of billions of years ago.

The oldest rocks now known on earth, as determined by radioisotope studies, are about 3.7 billion years old. Yet there are good reasons for thinking that earth took on her present solid figure well before this time. Certain studies of the ratio between the amount of lead that has been produced by uranium decay and the so-called primordial lead that was originally incorporated in earth suggest an age of 4 to 5 billion years. The younger ages for crustal rocks are explained by assuming that 3.7 billion years ago a remelting set all the radioactive clocks running anew.

Since astronomers believe that the entire solar system was formed at about the same time, give or take a few tens of millions of years, it would be interesting to get independent estimates for the age of other objects in the solar system.

The first extraterrestrial objects we were able to get our hands on were meteorites, chunks of stone and metal from outer space that heat up as they whizz through the atmosphere to produce flaming meteor trails. Some of them apparently come from the asteroid belt, a jostling ring of microplanets and smaller rubble that lies between the planetary orbits of Mars and Jupiter. This is believed to have been produced by the breakup of a planet, perhaps together with a family of satellites. Other meteorites may represent a phase in the condensation of the primary solar nebula when many aggregates did not yet exceed fist size. Radioisotope dates for meteorites lie between 4 and 5 billion years ago.

With some difficulty we have been able to get moon rocks into the laboratory. The oldest of these are at about 4.5 billion years B.P.

A maximum age for the earth of 5 billion years therefore seems reasonable. The very existence of potassium 40 (half-life, 1.3 billion years) indicates that the earth cannot be more than 6 or 7 billion years old; otherwise the amounts of this radioisotope would be essentially zero.

Theories and fashions change rapidly in astronomy, but the consensus seems to be that the oldest stars in our galaxy are 11 billion years old. The red shift of light from distant galaxies has been interpreted as meaning that the universe originated from a common center 15 billion years ago. Since it may be that this center, usually referred to as a "fireball" and consisting mostly of energy, originated from the collapse of a preexisting universe of galaxies, we are left with a picture of a universe that is without beginning or end, a pulsating glow that brightens and fades and brightens again for all eternity.

Earthquakes and Other Disturbances

More compelling than hot springs and upwelling lava as evidence of fierce energies latent within earth are the appalling shocks of heavy earthquakes and volcanic explosions. Probably the most powerful known earthquake was the one that destroyed Lisbon. In this Portuguese city, at 9:30 on the morning of All Saints' Day, November 1, 1755, a hammerblow from beneath turned the crowded stone churches into rubble. Two more shocks followed, and in 9 minutes the city was demolished, hidden in clouds of choking dust. A wave from the sea 40 feet (12 m) high poured over the harbor area, flames started by domestic fires swept the rest of the city, and 60,000 people died. It has been estimated that this quake had an intensity of 8.7 or more on the Richter scale. A powerful earthquake in Tibet in 1950 had a rating of 8.6 and was estimated to have released energy equivalent to the explosion of 100 fusion ("hydrogen") bombs.

Over the time since good records have been kept there have been 10 to 15 major earthquakes yearly (registering 7 or more on the scale). Quite possibly the annual world production of atomic explosives has about the same potential for energy release as earth's yearly production of energy by big earthquakes.

Energy released by earthquakes is only $1/8$ that yielded by volcanic eruptions, on a worldwide basis. However, the heat energy from volcanoes is about 10,000 times as great as the explosive energy they produce. The most remarkable volcanic explosion of modern times was that of Krakatoa, a volcanic island lying in the narrow strait between Java and Sumatra. In May of 1883, after a few years of premonitory earthquakes, the volcano began to pour out pumice and dust, along with a barrage of minor detonations. Then late in August began a series of mighty explosions. Those of the morning of August 27 were heard as far as 3,000 miles (4,800 km) away. A hundred miles (160 km) away, lamps had to be lit during the day because of the heavy cloud of dust, and for years afterward the atmosphere gave unusually reddened sunsets caused by the haze of ash still aloft. One estimate gives the largest single blast force as equivalent to about half the force of a hydrogen bomb. The total energy yield during the entire eruption was estimated at that of 10,000 such bombs.

Apart from studies of the kinds of damage to buildings and observations of cracks and ground displacement, precise studies of earthquakes did not begin until late in the nineteenth century. The center from which the study of earthquakes grew was Japan, where the foreign scientists who had been imported to accelerate the development of Japanese science were intrigued by the numerous quakes, large and small, that were an important part of Japanese life. Some of the first accurate data on surface phenomena were gathered here. In the Valley of Neo earthquake of 1891, a displacement of the ground crossed the island of Hondo—the vertical slip was measured at 23 feet (7 m). The deadliest was the Kwanto earthquake of September 1, 1923 in which large areas of Tokyo and Yokohama were devastated. Casualties were enormous, with a quarter of a million fatalities, but most were caused by fires that swept the flimsy bamboo and paper dwellings. The center of the quake was at Sagami Bay. Studies of the floor of the bay seemed to indicate remarkable changes, with some areas lowered as much as 750 feet (230 m), others raised by 885 feet (270 m). The ferocity of the quake lent credence to these figures, but later it was believed that these changes were partly the result of settling of fluffy bottom sediments derived from nearby volcanoes, of mud slides, and of inaccurate maps of the harbor made before the quake. Careful measurements on dry land indicated vertical displacement of about 6.5 feet (2 m), horizontal nearly 10 feet (3 m).

An Alaskan earthquake at Yakutat Bay in September of 1899 produced a record ground movement, a vertical slip of 52 feet (16 m). The great San Francisco quake of 1906 was caused by a horizontal slip along the San Andreas fault, which parallels the coastline some distance inland. The greatest known slip along the fault coincident with the 1906 quake was 23 feet (7 m). Studies along this fault show that the total displacement over the past has been at least 300 miles (500 km), and is probably much greater than this. All indications are that the part of California west of the San Andreas is on its way northward and in some millions of years will lie off the western coast of Canada.

The immediate cause of nearly all earthquakes is movement along already established breaks in the earth's crust, the geologic faults. There are countless faults in the rocks, most of them long quiet. Others, usually recent in origin, are the site of almost continuous quakes (mostly very small), with destructive tremors occurring only at long intervals. The quakes that leave traces on the more ordinary seismographs number about 100,000 a year. It is evident that the underlying forces that cause movement along the faults do not operate in the same violent and convulsive fashion as the quakes themselves. Rather, the force is continuous, bending the sheets of rock that lie on each side of the fault and store up elastic energy, as in a hunter's bow. Friction between the rocks at the fault surfaces tends to keep them from moving, but at irregular intervals (since the amount of friction depends on a variety of factors) the resistance is overcome, and the rock faces of the fault suddenly slide past each other.

A useful by-product of earthquakes is a picture of the planet's internal anatomy. Earth movements produced by earthquakes cause two kinds of waves. These are compression waves, which travel through both solids and liquids, and shear waves, which travel only through solids. Ground motion caused by compression waves is on the line passing through the quake center and the recording instrument, that caused by shear waves is at right angles to it. Thus, it is easy to distinguish between the two on the recording instrument, a seismograph.

After a powerful earthquake, the shocks can be detected on the opposite side of the globe, the compression waves arriving in about 22 minutes. In a large circle directly opposite the quake, only compression waves occur. In a wide band around this circle, like the iris of an eye around the pupil, no strong waves of either type are received. Outside this are both compression and shear waves, which are strongest near the quake. This eyelike pattern can be explained by assuming that the earth has a liquid core approximately 4,300 miles (7,000 km) in diameter. Being liquid, it completely blocks the shear waves. Since it is a sphere, it acts as a lens and focuses the compression waves into the "pupil" area, leaving a shadow in the iris. It must be assumed that outside the liquid core is a thick blanket of material, extending to the surface, that transmits both compression and shear waves.

It is believed that the core of the earth is mainly liquid iron (modern theory calls for an innermost solid fraction of the core), which together with the relatively light materials around it would bring the density up to the required 5.48. Most of the meteorites that are found (not a fair sample, since the stony ones are harder to detect among ordinary rocks) are composed of iron or iron alloys. These metallic meteorites are thought to include fragments of a planet that may once have orbited in the too-wide gap between Mars and Jupiter, lending support to the idea that the earth contains huge quantities of iron and iron-nickel alloys.

Early in the twentieth century an analysis of shock waves within about 125 miles (200 km) of a quake center showed certain anomalies that could be explained by assuming that some 20–25 miles (35–40 km) below the surface the material of the earth suddenly became more dense. This boundary was termed the Mohorovičić discontinuity, after its Yugoslav discoverer. Above the "Moho," as it came to be called, was the crust; below, extending all the way to the central metallic core, was the very thick mantle. When it was studied on a world-wide scale, the crust was observed to vary in thickness. Under the deep oceans, it was about 6 miles (10 km) thick. At the site of the great mountain ranges, it might be as much as 44 miles (70 km) thick.

The Moho was not a smooth boundary. Where a mountain range towered into the air, the Moho descended into the mantle; the mountain range had "roots." Under the oceans, the Moho was closer to the surface than under the continents. It was as if the lighter crust were floating on a liquid mantle. Some tens of thousands of years ago Scandinavia was covered with a layer of ice several miles thick, as is Greenland today. The Scandinavian ice sheet has since melted away and, as the "floating" theory requires, the land surface has rebounded. This is shown by a succession of ancient beach terraces, the highest being 1,650 feet (500 m) above sea level. Calculations show that the rebound has not yet been completed and that the land has another 650 feet (200 m) to rise to make up for the lost weight of the ice. This lag, together with other observations, suggests that the mantle is not a true liquid or, to put it another way, is an extremely viscous liquid.

Exposed to the sudden shock of earthquakes, the mantle behaves like a solid. Under the unrelenting, multimillennial pressures of the overlying crust, it behaves like a liquid. Window glass is a similar substance. In our ordinary experience it is a brittle solid, but in old buildings a window pane is thicker at the bottom than at the top, since over the years it has sagged like soft wax.

With the hypothesis (or theory) of a liquid mantle, it became possible to revive the old theory of convection currents in the deep interior of the earth as an explanation for various geological phenomena, to the point that it came to occupy a central position in modern theoretical geology.

The idea that the continents have not always been in their modern positions is chiefly the conception of the German meteorologist and geologist Alfred Wegener (1880–1930), who promoted the idea of "continental drift" from about 1912 to 1928. During his lifetime it was a minority theory. One standard American textbook of geology published in 1931 said that there was probably no truth in it. In addition to such obvious evidence as the fit of the western bulge of Africa into the depression holding the mouth of the Amazon River, Wegener adduced a great amount of evidence from the distribution of living and fossil plants and animals. Some of the fishes of South America are related to those of Africa, for example, and are found today in neither North America nor Europe. Since these fishes (lungfishes) live in fresh water, and presumably were unable to cross oceans, the two continents must once have been joined. However, fossils ancestral to both African and South American lungfishes occur in ancient rocks of North America. This evidence is of course not decisive; it allows the Wegener hypothesis to be either true or untrue. In sedimentary rocks about 300 million years old, of the Paleozoic era, there were found in South America, Africa, and Australia (and, most recently, in the Antarctic) abundant fossil leaves of an extinct seed-fern called *Glossopteris* because of their distinctive tonguelike shape. In rocks of corresponding age in the northern continents the abundant plant fossils, which have produced the anthracite coal beds of the eastern United States, are of entirely different types. This ancient flora of the Southern Hemisphere supports the idea that the southern continents were once a single land mass.

During the 1960s a detailed study was made of the

Era	Period	Epoch	Time since beginning (millions of years)
CENOZOIC	Quaternary	Recent	.01
		Pleistocene	2
	Tertiary	Pliocene	13
		Miocene	25
		Oligocene	36
		Eocene	58
		Paleocene	65
MESOZOIC	Cretaceous		136
	Jurassic		190
	Triassic		225
PALEOZOIC	Permian		280
	Carboniferous		345
	Devonian		405
	Silurian		425
	Ordovician		500
	Cambrian		600
PRECAMBRIAN			4,000

geology of the African bulge and the South American depression, using radiometric methods for determining ages as well as standard methods of determining mineralogical content. An exceedingly close match of the rocks on both coastlines was found. So far as is known, the rocks on the ocean floor in between are no older than Cretaceous (135 million years B.P.) or Jurassic (180 million), and are for the most part younger than this, while those of the African and South American coasts are Cambrian, between 500 and 600 million years old. This and other evidence seems to prove beyond reasonable doubt that the two continents once were part of a single land mass.

Another seemingly unrelated kind of evidence also came to bear on the problem of continental drift. It has long been known that igneous rocks such as lava are weak permanent magnets. When the molten rock pours out on the surface and slows to a halt just before congealing, the atoms of iron, still mobile, orient themselves in a north-south direction like so many compass needles. Then as the rock cools down to some few hundreds of degrees Centigrade, they are frozen into position, making the rock a magnet.

Study of these fossil compass needles is now a flourishing subscience of geology. Like the study of radioisotopes it is a difficult and complicated art, with many pitfalls, but by dint of hard work and much trial and error it has been developed into a useful tool for studying the history of the earth. It seems that the ancient compasses pointed in different directions in different continents at, as near as can be determined, the same time. This could be explained if the earth had more than one pair of magnetic poles. A more reasonable explanation is that the continents themselves have since drifted and become reoriented. Paleomagnetic studies in conjunction with other evidence suggest that North America and Eurasia separated at about 60 million years ago, South America and Africa 135 million, and Australia and Antarctica 50 million years ago.

Geology, which seemed to be a science that was growing old gracefully, underwent a revolution in the 1960s like that in physics when radioactivity was discovered and in biology when the chemistry of the gene was revealed. The subversive influence was study of the floor of the open seas.

Although when ships are near a coastline they have to take soundings to keep from running aground, the depth of the water on the high seas is of only academic interest. The first really deep sounding, one of 21,600 feet (6,580 m) was made in 1839 with a weighted rope that had to be specially constructed for the job. It might take a whole day to make a sounding like this, and by the middle of the nineteenth century only 180 deep soundings in the Atlantic had been made. During World War II a device called sonar was perfected that sensed solid objects under water by echos and was used mainly in locating submarines. After the war it was used to make topographic maps of the sea floor. A cruising ship could easily make a continuous record of the distance to the sea floor beneath.

There gradually took shape on the maps the details of a gigantic mountain chain that threaded its way down the middle of the Atlantic from the Arctic almost to Antarctica. For most of its length it lies midway between the coasts of the Old World and the New. This Mid-Atlantic Ridge goes through Iceland and brings to the surface the Azores, the St. Paul Rocks, Ascension Island, the Tristan da Cunha group, Gough Island, and Bouvet Island. When the age of the rocks of these islands is measured radiometrically, the surprising fact emerges that they are very young, only a few million years old. It had been basic doctrine that the ocean floor was the most ancient of the main features of the earth and was covered with thick sediments that had been accumulating for billions of years. The youth of the islands did not in itself contradict this, since they might be the caps of recent volcanic ranges that had thrust themselves up through the ancient floor. However, seismic sounding methods of mapping rock layers, like those used by oil geologists on dry land, and drill cores made short distances into the sea floor confirmed that the floor near mid-ocean was young. Here there was only a superficial layer of fine sediment, of recent age and derived mostly from the skeletons of marine microorganisms, covering basalt (hardened lava).

When the ages of the rocks of the mid-Atlantic islands were studied in detail, it was discovered that the age was proportional to the distance from a crack down the center of the ridge, called the Rift. The farther from the Rift, the older the rocks. To account for this, it was proposed that new sea floor was constantly being made by lava flowing up through the Rift, heaping up to form the mountainous ridge, congealing, then being pushed aside in each direction, carrying the islands with it. The ridge is an active center of earthquakes, and there are active volcanoes where it goes through Iceland. The islands were carried east and west of the Rift to distances that indicated a rate of drift of about an inch (2.5 cm) a year. In Iceland, measurements made on the ground in the Rift over a period of years showed markers separating at the rate of a quarter of an inch (6 mm) a year. The same mechanism was used to explain the drift of Eurasia and Africa from North and South America. An inch a year in each direction from the Rift would carry the eastern tip of Brazil to its present 3,000 miles (4,800 km) from the Gulf of Guinea in about 100 million years.

The oceanic islands in the Atlantic are few, and the fact that their ages support the theory of sea-floor spreading from the Mid-Atlantic Ridge might be only coincidence. What was needed was a method for determining at least the relative ages of rocks deep under the sea. A most improbable development in the study of paleomagnetism made this possible. It was found that some rocks, so young that they could not have been shifted to any great degree by continental drift, had magnetic fields that were reversed, as if they had been formed when the magnetic poles of the earth itself had been reversed. Nothing in theoretical geophysics had predicted anything like this, but once discovered, it was not so bad after all—minor shifts in currents within the core, requiring only small inputs of energy, apparently could result in a rather abrupt reversal of the field.

One set of investigations disclosed that there have been five periods during the past 3.5 million years when the magnetic poles were reversed. One of these periods lasted nearly a million years, and in all, only about half the time was the polarity normal—that is, as it is now.

While this work was going on, geologic surveys had begun to study magnetic anomalies from the air as a modern method of prospecting, since these often show the presence of large ore bodies containing metals. An airplane tows a sophisticated sensing device that records slight deviations from the general magnetic field of the earth. In more academic research, it was found that these deviations could also be found over the oceans, and extensive oceanographic surveys were made.

Over the Atlantic, a remarkable symmetric pattern of bands indicating abnormalities appeared on the maps, paralleling both sides of the Mid-Atlantic Ridge, a pattern like that seen in following a straight line across a tree stump, with a balanced series of tree rings on each side of the center. Each of the magnetic bands is caused by the reversed magnetism of the rocks of the sea floor. The fact that the bands, which are recognizable by their unique width, occur in pairs equidistant from the ridge is explained if they were once a single band at the ridge, then split and were pushed apart by sea-floor spreading. The mechanism is like a giant tape recorder, the molten rock acting as the tape and recording the message—the direction of the earth's magnetic field—as it congeals.

The immense quantities of molten rock required for sea-floor spreading cannot be delivered to the oceanic ridges without some kind of circulation in the mantle. Usually it is assumed that an oceanic ridge is the line where a convection current rises to the surface, then splits left and right and pushes the crust in both directions by forcibly injecting rock into the rift and by drag on the underside of the crust. If there is upward flow at the ridges, there must be downward flow somewhere else, since the earth's size remains the same. If crust is being formed, it also has to be destroyed.

Although this cannot be observed directly, it is hypothesized that there is downward movement where two blocks of moving crust, called tectonic plates, collide. The collision fronts are of two types: (1) trenches, in which one plate slips under the other, plunging down into the mantle, and (2) huge young mountain ranges, where the plates meet head on, buckling upward. Trenches are believed to be formed when the plates are moving fast, more than 2.4 inches (6 cm) per year. At the collision front, a line of great volcanoes is formed. A good example of such a front is the great arc of volcanoes running through Japan, the Philippines, and New Guinea, which face deep oceanic trenches. Mountain-building collision fronts are presumably formed when the plates move more slowly. Examples are the Andes of the west coast of South America; the Himalayas, where the plate carrying India crashed into southern Asia; and the Alps, where the Africa-bearing plate collided with Europe.

Both the trenches and the mountain-building lines of collision are believed to be places where materials are returned from the crust to the mantle. Under the great mountain ranges along continental margins are the deep roots of the mountains, analogous to the underwater part of the floating iceberg. As these roots grow from materials added by the descending plate, they are melted by the hot upper layer of the mantle.

The vast majority of the world's earthquakes occur

along the two great areas of tectonic plate collision: the western coasts of the Americas, and the complex system of trenches and volcanoes resulting from the collision of three tectonic plates in the western Pacific. The Alps and Himalayas are thought to result from the collision of tectonic plates that all carried continents. These ranges also are active sources of quakes. By contrast, the quiet Alleghenies are being smoothly carried along on the tectonic plate that originates in the mid-Atlantic. It is only the western third of the North American continent that is crumpling under the impact of this tectonic plate with one originating in the eastern Pacific.

Although there can be little doubt that huge blocks of the earth's crust, carrying whole continents, do move, and are created at one edge and destroyed at another, the underlying mechanism has not yet been satisfactorily explained. No model of the earth, complete with moving parts, has been constructed that can explain the size and distribution of the continents, for example. Even the origin of the continents, with their thick light crust, remains a mystery.

The earth is a living planet not only in the sense that it bears plants and animals but also in the sense that it is everchanging, in cycles that are ages or hours long, and in slow irreversible developments like those of a growing and aging man. The axial spin gives day and night, the inclination of the axis gives the seasons as the planet orbits about the sun. The spin of the earth represents an enormous amount of kinetic energy. If it were completely without friction, the spin would last forever; its endurance for billions of years shows that it is very nearly frictionless, considering the great amount of energy involved. The fact that it is not entirely frictionless, however (which has gradually made the days longer), has important consequences for the earth. Part of the loss of kinetic energy is caused by the tides, both the massive wash of the oceans and the almost imperceptible rock tides, which bend the crust of the earth twice each day. This kind of friction results from the existence of our huge satellite, the moon, and the sun. But the bulk of the oceans is infinitesimal compared to that of the rest of the earth.

Direct measurements show that the rate of spin varies slightly from day to day, season, and year to year. This introduces a "rumble" or very slight axial disturbance into the smooth spin of the earth, like a bad bearing in an engine. Since the core is liquid, the application of this energy to the interior would be expected to keep currents moving, especially if periodicity and resonance are involved. Resonance could presumably keep even the viscous mantle circulating. There is good evidence that the change in rate of daily rotation is related to earthquake activity. Even these small changes in inertial energy would be enough to raise the Himalayan mountain system from sea level during a time span on the order of a million years.

Aside from radioactive heating and loss of the heat remaining from the earth's creation, these irregularities of the earth's spin are the only known source of energy for moving the contents of the mantle and core.

The existence of the continents has not yet been explained satisfactorily. They are made mostly of light granite in contrast to the sea floor, which is mainly heavy basalt. One suggestion is that when tectonic plates collide, the one that descends begins to melt at a certain depth, and the lighter components of the basalt rise to the top, where they are skimmed off on the edge of the other plate and heaped up on its surface. At the western margin of South America, for example, the lightweight and light-colored semiliquid rock that will eventually become granite is scraped off like cake frosting as the Pacific Plate comes up against the continental margin and slides under it. Perhaps billions of years of spread and collision of tectonic plates have gradually built up the continents in this way, skimming off and massing together the lightest upper layers of the oceanic crust.

In the sea are dissolved a great variety of chemicals that make it a rich mineral soup. The internal juices of the animals that live in the sea, even the blood of animals like ourselves whose ultimate ancestry was in the sea, have a chemical kinship with this primordial brew. If it were not for the continents, the composition of the ocean would be much different. Rains leach out soluble minerals from the rocks and soils of the land, rivers carry them to the sea, and evaporation leaves them there. In the slow rise and fall of the continents, shallow bays are formed, then isolated, and the oceanic salts are deposited as thick layers of minerals when the bays dry up. Thus, over the billions of years, a balance has been struck in the ocean between gain and loss of minerals.

Most kinds of animals and plants live on the land. The land's physical habitats are far more diverse than the sea's, giving evolution the opportunity to fill a greater variety of habitats with organisms adapted to these local and varied situations. Being close to unlimited supplies of the minerals needed to build and operate living tissue, life on the land is more abundant, per square mile, than under an average square mile of ocean. The reason is that although the elements exist in great variety in the sea, some are so dilute that they limit the amount of life that the ocean can support. The existence of land has had a most profound influence on the course of development of life on the earth.

The mere fact that the continents are separate blocks of land, sometimes completely disjoined, sometimes linked by narrow strips of land, has given impetus to evolutionary advance and to diversification. In freedom to experiment, different kinds of animals evolve, by chance, on the different land masses, isolated from interference from outsiders. When the continents are joined, by uplift of land or lowering of sea level that exposes land bridges, or by collision from continental drift, the animals and plants mingle, with the fortunate species surviving in the competition, the rest departing into oblivion. As the continents have slowly drifted from cold regions to hot, then from the tropics again into high latitudes, the inhabitants have met new challenges and have had to adapt.

Even before the appearance of living plants and animals, the earth was an active bit of astronomical real estate. As we shall see in a later account of biological evolution, without the challenge of a dynamic, changing environment the evolution of life above a very primitive level would probably have been impossible.

However, it is possible to have too much of a good thing. Should challenges be too severe, and the changes take place too rapidly, organisms might not be able to adapt. The instability of the earth's axis (which is believed to cause most of the geological disturbances of the kind discussed), as well as climatic changes, is quite moderate when compared to another earthlike planet, Mars. It is believed that the nearby presence of the moon stabilizes the earth's axis. Also, it is believed that the moon contributes to the strength of the earth's magnetic field, which partially shields it from the solar wind of high-energy particles. No other planet has a nearby massive satellite, and one would imagine that they would not be common in other solar systems.

It can be seen, from this brief look at the anatomy and physiology of the earth, that the sun and our planet may be rare birds. To say that the suitability of the earth for living things is remarkable is something like remarking how fortunate it is that San Francisco or New York has a good harbor. But it may be that good harbors for life in the galaxy are few.

THREE

•

THE BIOSPHERE

EARTH IS INHABITED by millions of kinds of elegantly designed plants and animals that chemistry and physics never dreamed of. These living things form a discontinuous layer, called the biosphere, of forest, grassland, numberless creatures scattered through the sea, and an evanescent and scanty desert biota. Were this spread evenly over the surface, it would make a layer about 4 inches (10 cm) thick. It would weigh about 20 million million tons.

Other superficial layers of the earth are the hydrosphere (oceans and fresh waters), which weighs about 1,400,000 million million tons, and the atmosphere, weighing 5,200 million million tons.

In spite of its small size, the layer of living substance is important in the affairs of the earth's surface. Probably nearly all of the present atmosphere has been produced by the chemical activity of the biosphere. The other activities of the biosphere are on a scale that brings it into the range of the major geologic forces that operate on the surface.

Each year the biosphere burns a total of 35 billion tons of carbon to furnish the energy for the manifold activities of the creatures that compose it.

Burning 35 billion tons of carbon a year in the metabolic machinery of living substance produces an energy yield which, if spread out evenly over the surface of the earth, would average between 1 and 2 one-millionths of a calorie per square centimeter ($\frac{1}{6}$ square inch) per second. This is almost the same as the flow rate of heat from the interior of Earth across its surface, and the energy involved is much greater than that of volcanic eruptions and earthquakes.

The energy for photosynthesis is provided by sunlight. Only about one-tenth of 1 percent of energy received from the sun by the earth is fixed by photosynthesis. This fuels essentially all activities of the biosphere except those based on the oxidation of fossil fuels.

The Origin of Life

So far as direct (or at least obvious) evidence from fossils takes us, the biosphere appeared fairly late in earth's history, about 600 million years ago. These fossils are of rather complicated animals, all marine, and large enough to be seen without a lens.

The study of fossils, paleontology, was essentially founded early in the nineteenth century by the efforts of two French scientists. Jean Baptiste Lamarck (1744–1829) can be regarded as the father of invertebrate paleontology (with books published in 1801, 1802, and 1809), which at first concerned itself mainly with the shells of snails, clams, and other molluscs. And in 1812 Georges Cuvier (1769–1832) founded vertebrate paleontology, the study of old bones. There seems to be no similarly spectacular beginning for paleobotany. Despite some 150 years of research, however, paleontology failed to clear up the mystery of the sudden appearance of abundant complex animals in Cambrian rocks at 600 million years B.P. (700 million, if we include the recently discovered Ediacaran fossils of Australia, but these make a poor showing).

According to evolutionary theory, life must have had a long period of development preceding these fossils. The continually fascinating questions are when and how life began and what forms it took. It is certain that our planet is older than the oldest radioisotope dates we have for earth rocks, which are placed at 3.7 billion years B.P. The consensus is that this limit results from a surface melt reaching about 8,000° F (4,500° C). Surface conditions and the atmosphere must have been altered so drastically that we can ignore earlier conditions in speaking of the environment in which life originated.

As the surface cooled, a solid crust formed, and immense quantities of water vapor that had been forced out of deep molten rocks by volcanic and plutonic activity condensed to form the seas. Gases from the molten rocks had produced an atmosphere believed to have consisted mainly of carbon monoxide, carbon dioxide, nitrogen, and hydrogen, together with water vapor.

The radioisotope clocks started with the solidification of minerals that contain radioactive elements. The presence of sedimentary rocks soon after the formation of a solid crust proves that there were continents, oceans, and atmosphere, that there were gradients and rainfall to wash minerals into the ocean. Details of the atmosphere's composition vary from one authority to the next, but all agree that oxygen was not present in significant amounts. Free oxygen has not been detected in the atmospheres of other planets. It has not been found among the gases coming from volcanoes.

These findings fit well with theories of the terrestrial origin of life, which require a prebiological history of the development of organic compounds. These cannot exist in the presence of the chemical ferocity of free oxygen. If we assumed an atmosphere containing oxygen, we would also have to assume that the original living microbes drifted onto the earth from outer space as inert spores, resistant to the intense cold and radiation of that environment but ready to spring to life when they reached favorable conditions. The latter is a respectable scientific hypothesis and has been considered by a number of investigators. The Swedish physical chemist Svante Arrhenius (1859–1927) thought such spores could move through the vast reaches of space driven by the pressure of light from stars.

Modern cosmologists tend to reject the theory of primitive life from outer space colonizing the earth. The main reason is that they believe the universe to be old enough for only three or four generations of solar systems to have existed. All that would be gained by the immigration theory would be to push the problem of the origin of life back some billions of years to some unknown place. Better to deal with the problem here, where more or less educated guesses can be made about conditions prevailing at the time when life had to originate.

We return to our vision of a lifeless earth almost 4 billion years ago. There are great oceans, presumably blue and sparkling. Waves wash the shores of continents. Clouds float in the sky. High cumuli reach up into freezing temperatures, producing ice nuclei that form raindrops. The atmosphere is a poisonous mixture of gases, rich in carbon monoxide and hydrogen. It could be bottled under pressure and used as a fuel like modern butane tanks.

Pouring into the atmosphere is a rain of high-energy radiation from the sun, including powerful ultraviolet rays that no longer penetrate the atmosphere because they are mostly absorbed by the oxygen in it. This radiation converts the atmosphere and the upper layer of the sea into a laboratory where organic compounds are synthesized from the simple compounds of the atmosphere, probably by the thousands of kinds and billions of tons per year as millions upon millions of years pass. In addition to energy from ultraviolet radiation, the chemical works probably get supplementary energy from lightning storms.

In 1953 Stanley Miller (1930–) brought the problem of the prebiotic production of organic compounds forcibly to the attention of both scientists and the general public by some well-designed, direct, and simple experiments. A mixture of gases then thought to represent the atmosphere of the primitive earth—methane, ammonia, water, and hydrogen—was held in a small glass sphere. Electric sparks were sent through it, in imitation of ancient lightning storms. In a week's time a number of organic compounds had accumulated in a collecting flask of water. Nineteen were identified, including two kinds of amino acids.

Many similar experiments have since been made, modifying the mixture of gases according to the latest theories of earth's history and using ultraviolet radiation instead of electrical sparks. All gave similar results. Pouring energy into artificial atmospheres or solutions containing water, hydrogen, nitrogen, and carbon monoxide or dioxide, together with a pinch of other simple compounds according to taste, produces a variety of complex carbon compounds, including amino acids. In solution, these in turn interact to produce even more variety and more complex molecules, especially if a little phosphorus is added. Over years of time the number of compounds apparently would be without practical limit.

For the first 100 or 200 million years after the great melt the mineral kingdom must have been very different from what it is now. Some few thousands of mineral species now exist. In this ancient prebiotic phase of earth's history the carbon compounds (which probably existed in hundreds of thousands of kinds in solution) would have to be called minerals in those instances where they existed as solids. We have no inkling of the abundance of crystals and blocks of organic compounds in the mineral kingdom as it existed just before the middle of the third eon B.P. Perhaps in desert regions there were dunes and plains of snowflakelike sands and wind-scoured pebbles of exotic carbon compounds, produced in dried-up sea basins, whose composition cannot even be imagined.

However that may be, we can be certain that the seas were a rich mixture of carbon compounds in solution.

Experiments of the Miller type show that the nonbiological origin of amino acids presents no problem. They probably existed in multi-billion-ton amounts long before the first living things appeared. But how did they come to be linked into chains, or polymerized, to form proteins? Although abundant in solution in the sea, they would not combine simply because they were there; the polymerization of amino acids is an uphill reaction, in the sense that it requires energy. If there were combinations of amino acids with other compounds that were energy rich, then one can imagine chains being formed, with the adjuvant molecules falling away and leaving the amino acid polymers (polypeptides) intact. However, no successful model for the spontaneous polymerization of amino acids in the prebiotic sea has been constructed in the laboratory.

If the amino acids are dry, the energy needed for their combination into chains is easily provided by heating them. Experiments of this kind were begun by the American biochemist Sidney Fox (1912–) in the late 1950s. In a typical example, he began with 10 g (0.35 ounces) of a single amino acid (say, glutamic acid) that melted at about 355° F (180° C). To this was added 10 g of another amino acid (aspartic acid) and finally 5 g (0.18 ounces) of a mixture of the remaining amino acids. This was kept under an inert atmosphere of nitrogen for 4 hours at 340° F (170° C). About 25 percent of the amino acids were converted into polypeptides.

Fox called these polypeptides "protenoids" rather than proteins, because proteins tend to have a highly specific structure. Some of the polypeptide chains he produced had molecular weights up to 10,000 (in this situation, the higher the molecular weight, the larger the molecule). Natural proteins (which also are polypeptides) have molecular weights as low as 10,000 and as high as 1 million. Arrangement of the amino acids was nonrandom in the artificial polypeptides. Some of the protenoids had catalytic properties (but not at the enzymatic level of efficiency) and showed a tendency to bind with heavy metals, which improved their catalytic properties.

It is obvious that the dry-land natural synthesis of proteinlike molecules offers difficulties in imagining their role in the evolution of primitive organisms. Fox speculates that organic compounds, including amino acids, would have been baked on warm or hot, newly formed volcanic rocks to form polypeptides, which were then washed by rains into rivers and the seas. This sequence does not seem overly contrived.

When Fox continued his experiments by adding the artificial protenoids to water, he confirmed what had already been known from experiments made early in this century with natural macromolecules such as proteins and gum arabic. The protenoids immediately formed spheres of bacterial size (2 microns, or 0.0078 inches, in diameter) which were bounded by a membrane. Some developed small buds that broke off when the suspension was shaken, then grew to full size. Fox called the structures produced in these experiments "protenoid spheres."

Large organic molecules such as polypeptides act much differently in water than do small soluble molecules. In fact, this branch of chemistry used to be called colloid chemistry, a term rarely heard in these days of molecular biology. The macromolecules are not said to be in solution, but in suspension. They can be sedimented out of the water by using a high-speed centrifuge. A colloidal suspension also differs from a true solution in that a beam of light passing through a suspension is visible when viewed from the side, whereas one through a true solution is not.

The size and electrostatic properties of macromolecules make them able to manipulate water molecules to form bounding membranes, and to maneuver the colloidal suspension within a great range of viscosities, ranging from Jello-like substances to thin liquids. The tendency of colloidal suspensions to round up spontaneously into droplets of relatively viscous, colloid-rich spheres suspended in liquid led the Russian scientist A. I. Oparin to study them in the early 1900s as possible intermediates between living systems and simple solutions. Colloid chemists called the microspheres "coacervates." Oparin wrote extensively on their lifelike properties.

Leading theorists of a few decades ago thought that the first living things were single molecules capable of reproducing themselves. The modern consensus is that they were wrong and that the first living things were microbelike structures such as the microspheres studied by Oparin and Fox. These structures at first grew rather uncertainly by the absorption of polypeptides through the surface membrane. There could be selective intake of other organic compounds which, in the small sheltered world provided by the membrane surrounding the droplet, could interact to provide energy for linking amino acids into polypeptides in place, instead of relying on the land-produced protenoids envisaged by Fox.

It may well be that at about 3.6 billion years B.P. there was a vast array of colloidal structures in the primitive sea. Some of them would have been the bacterialike coacervate droplets or protenoid spheres. Others might have been sheets and slabs of colloid-rich gels on rocks and in the muds of the sea floor. Perhaps modern biologists transported to the scene by time machine would argue endlessly over which were alive and which not. The boundary between living and nonliving was diffuse, not sharp. We could not say that in this particular particle the spark of life first appeared and burst into a steady flame. Perhaps, after a century of laboratory experimentation, biologists will look back and say that the experiments of Sidney Fox marked the first synthesis of life.

Early Life

We would imagine the first living things to have been microscopically small, soft blobs of matter resembling modern bacteria. The transitional forms would also have been small. From what we know about earth's early history, during more than half the life of the planet the continents were scoured with a deadly blast of ultraviolet radiation from the sun, which would prevent life from invading the land in any but local and protected situations. Organisms living in the water can remain soft and jellylike, not requiring hard supporting structures. We would expect that their remains, falling to the ocean floor, would quickly decay without leaving a trace in the fossil record. Only after a long history of prey and predator would there be produced the sheltering armor and the piercing teeth and fangs that make elegant fossils.

In very old Precambrian rocks there are peculiar dome-shaped structures called stromatolites, sometimes several yards in diameter, which resemble rather uncommon reefs found on certain tropical coasts. These reefs are not composed of the skeletons of corals, but rather of materials deposited by a primitive kind of microorganism called blue-green algae. Since these microbes have been found to be quite different from other algae, they are probably best called simply blue-greens.

In association with some of these Precambrian stromatolites are layers and nodules of flint or chert, a translucent, hard rock with a waxy luster. These rocks are varieties of chalcedony, others being carnelian, agate, and jasper. Chalcedony in turn is a kind of quartz, a very common mineral with the chemical name silicon dioxide. It is deposited slowly from underground waters and is able to fossilize organisms with remarkable fidelity. Petrified wood often is made of chalcedony and, when cut into thin sections, shows under the microscope the minute details of cellular structure. In the mid-1960s the American geologist E. S. Barghoorn (1915–) and his associates began to study the Precambrian chalcedonies. They discovered that at many localities the rocks were filled with various blue-green-like or bacterialike microstructures. These at first seemed to show variety of shape and content that permitted classification into various groups, but experiments in which modern microorganisms were artificially preserved led the investigators to be cautious about drawing information from these ancient fossils. It seems probable that they are the remains of simple organisms. There is doubt about their degree of resemblance when living to the many kinds of microbes that now exist, but for convenience, and since they appear to have been photosynthetics, they will be referred to here as "blue-greens."

Another early indication of the presence of life is the existence of a peculiar rock that has stripes (horizontal where undisturbed) of black or brown iron oxide alternating with white quartz. This rock, found at many levels in the Precambrian, is called the Banded Iron Formation, or BIF. It is important because the iron oxide (magnetite) is believed to indicate the locally abundant presence of free oxygen. This free oxygen is believed to have reacted with greenish ferrous iron compounds to produce the more oxidized magnetite of the BIF.

Banded iron rock occurs in huge amounts, by the tens of millions of tons, and accounts for most of the world's reserves of iron ore.

The free oxygen probably was produced by the world's first abundant photosynthetics, or blue-greens. This theory is supported by the fact that stromatolites are associated with BIF.

Early in earth's history the impact of solar radiation on water molecules in the atmosphere probably split them into hydrogen and oxygen. But so overwhelmingly abundant were the reduced compounds in the atmosphere (those compounds which combine readily with oxygen) that the oxygen was removed chemically. At the site of BIF activity, it has been supposed, there was an oscillating balance. A surplus of oxygen alternated with scarcity as populations of blue-greens waxed and waned, producing banding in the sediments.

Later in geological history, beginning about 2 billion years B.P., the Banded Iron Formations give way to thick red beds, indicating a stable and overwhelming abundance of oxygen. This is assumed to represent a final victory for the photosynthetic microbes, perhaps as the result of a series of biological inventions that improved the metabolic machinery of these pioneers.

Several decades ago, the student of general biology learned that there were two main groups or kingdoms of organisms, the animals and plants. Further, within each of these kingdoms were a group of simple, single-celled organisms and a contrasting group of larger many-celled organisms that were aggregates of single cells differentiated for various functions. Today, the student learns that there are two basic kinds of cells: The simpler is without a nucleus and without chromosomes. This is the prokaryote cell, and the organism is called prokaryote. All prokaryotes are single-celled plants or animals, although some aggregate to form rude colonies of more or less similar cells. The more complex eukaryote cell has a nucleus in which are found a number of chromosomes rich in DNA (the prokaryote cell also has DNA, but microscopically this appears diffuse rather than concentrated into chromosomes). Many eukaryotes are single-celled, and all many-celled animals and plants (Metazoa, Metaphyta) are composed of eukaryote cells.

The living prokaryotes comprise two groups of organisms: the bacteria and the blue-greens. A few bacteria The more complex eukaryote cell has a nucleus in which are photosynthetic, that is, they can use sunlight as an energy source for their metabolism. However, they do not use this energy to obtain hydrogen from the water molecule. Instead they get the hydrogen, to use as fuel, from organic compounds or from hydrogen sulfide or other sulfur compounds. Thus they do not produce free oxygen. In contrast, blue-greens use sunlight to separate water into hydrogen and oxygen. They also have the ability to use nitrogen from the air to form amino acids. Thus they are remarkably independent (technically, they are termed autotrophs) and can live on air, water, sunshine, and a few minerals.

Comparatively little is known, however, of the details of the metabolism of blue-greens because they are not widely used in the laboratory, in comparison with bacteria. Also, it is virtually certain that bacteria preceded blue-greens in the history of life. Therefore we shall discuss the basic elements of biological machinery using a bacterium as an example.

The bacterium *Escherichia coli* (the term is used so often that the genus is usually abbreviated, *E. coli*) is the most intensively studied organism in biological science. Life is chemistry, and in this immobile but chemically active microbe the essentials of life are shown with clarity. It is estimated that about 3,000 chemical reactions, nearly all involving organic compounds, take place in this bacterium. About 1,000 of them are known. The reactions have the function of synthesizing new bacterial substance from the nutrients that are available.

The metabolic pattern of *E. coli* is basically the same as that of other living cells, in organisms both great and small. Some other organisms, to be sure, have added on whole arrays of patterns. In green plants, the apparatus for photosynthesis has been adjoined to the basic pattern.

In animals such as ourselves, the transformation of chemical energy into the mechanical energy of muscles makes an addition that is spectacular in appearance, and the special activities of the nerve cells allow spectacular achievements.

A most remarkable feature of these bacteria, and of all living cells, is one that separates them by a wide gulf from the rest of terrestrial creation. They possess thousands of giant molecules that are micro-galaxies of many thousands of atoms, each spaced precisely in complex, varied patterns in three-dimensional space. Most of the atoms are of hydrogen, oxygen, carbon, and nitrogen; among the others are some of the heavy metals. Each of these giant molecules (macromolecules) is a crystal, rivaling in complexity anything found in the mineral world. These macromolecules are the proteins. An individual cell of *E. coli* is 60–80 percent water. Of the remainder, about 70 percent by weight is composed of proteins. In a bacterial cell there are 3,000–6,000 kinds of proteins, about half of which are enzymes.

Enzymes are proteins that facilitate most, if not all, of the chemical reactions going on in a living cell. Thousands of kinds of enzymes are each a specialist for a different chemical reaction. This specialization is the result of the particular arrangement of atoms within the giant protein molecule. The enzyme molecule quickly fixes into place (at the "active site") the chemicals that are to undergo reaction, one molecule or pair of molecules at a time. The reaction products are quickly released, and a new reactant fixed in place. Different enzymes in the cell are linked in that they accept the reaction products of a different kind of enzyme, and pass on their own product to yet another. Metabolism of a cell can be viewed as a kind of assembly line (in reality not linear, but with many cyclical events and branching alternative sidelines) in which the components to be handled are individual molecules of small or moderate size.

Enzymes only facilitate reactions that can take place without them. The difference is that the enzymatic reaction proceeds at much greater speed. The biological chemical factory operates in cool water and has reaction speeds that are thousands of times faster than the same reactions show in the laboratory without the aid of enzymes.

When the nature of enzymes was comprehended, they seemed to possess the vital, magical force that was believed, by a school of biologists called vitalists, to exist in living systems but not in physical or chemical ones. However, it is possible to extract specific enzymes from bacteria. In the laboratory, in solutions with suitable properties, many enzymes will carry out the appropriate reaction with speeds like those found in the living bacterium. Enzymes have also been taken apart piece by piece, diagrammed, and then put back together with nothing but chemistry and the hard work and ingenuity of the scientist.

Many substances other than enzymes will accelerate chemical reactions without themselves being used up. These are termed catalysts. An example is platinum. Left to itself, a mixture of hydrogen and oxygen will not react. But if a polished slab of platinum metal is introduced into the mixture, the two elements combine to form water. Platinum is also used as a catalyst in automobile exhaust systems to bring about the combination of oxygen with substances that were not burned in the engine cylinders. Enzymes are catalysts, although they should be termed supercatalysts for their extraordinary efficiency, which is based on the highly specialized arrangement of atoms in the enzyme molecule.

An enzyme molecule, like other proteins, is basically a chain of small nitrogen-containing organic molecules called amino acids. There are about 20 kinds of amino acids, and the order in which they are arranged on the chain determines the properties of the protein or the catalyst. As it is being formed, the protein chain folds up into a three-dimensional pattern that is characteristic of the protein and is determined by the amino acid sequence.

Escherichia coli and other living cells contain the metabolic machinery that can assemble amino acids (which are either obtained directly from or manufactured from its food supply) into the proper sequence to make the thousands of kinds of proteins, including enzymes, that form a major part of its substance. The "blueprint" for this mammoth construction job makes up about 20 percent by weight of the nonaqueous portion of the bacterium. It is chemically known as deoxyribonucleic acid or, more familiarly, as DNA. DNA is essentially a chain of small nitrogen- and phosphorus-containing organic molecules called nucleotides. There are four kinds of nucleotides, and the information of the blueprint resides in their sequence. There are of course too few kinds of them to map one-to-one with the 20 amino acids, but there are more than enough combinations of three nucleotides taken at a time to have a triplet signify a particular amino acid. The machinery whereby this blueprint or code is realized in the final protein is fantastically complex, and up to now only the skeleton of the apparatus has been revealed.

In *E. coli* the DNA chain forms a loop which, being hundreds of times longer than the bacterium, is folded up inside the organism. Except for rare deviations that cause mutations, it is exactly replicated each time the cell divides. It thus constitutes the basic part of the cell machinery responsible for inheritance.

Living things produced enzymes by a long process of natural selection. The crude catalysis of relatively nondescript polypeptides gradually evolved into the precise architecture of enzymes. This could not take place without the coevolution of a polynucleotide system. Perhaps before the evolution of a fixed nucleotide pattern there was a diffuse high-energy phosphate system like that of the modern adenosine triphosphate (ATP) system that linked amino acids within the cell. First with clusters of nucleotides, then with larger chains, the blueprint for an entire enzyme protein molecule could be laid out. The modern conception is that the DNA-enzyme basis for the existence of all known modern living cells is the product of a long evolutionary history of living cells that were at first without it.

Modern organisms themselves give evidence that in their remotest ancestry they lived without oxygen. Many bacteria, called anaerobes, can live without oxygen. Other bacteria, and other microbes as well, can live either with or without oxygen. As was shown by Louis Pasteur (1822–1895), this is exactly what yeasts do. Without oxygen, they grow slowly, use nutrients inefficiently, and leave unused considerable amounts of energy-rich ethyl alcohol. Given enough oxygen, they grow quickly and leave only carbon dioxide and water.

In all aerobic (oxygen-using) organisms, there is a primitive anaerobic system for generating the energy needed for life activities, on which is superimposed a more elaborate system that uses oxygen to burn nutrients to carbon dioxide and water, thus extracting all their potential energy. A well-trained athlete can run for an hour or two at a rate of about 300 meters per minute. If he sprints, however, his leg muscles quickly become incapable of movement. The reason is that the circulatory system can not deliver oxygen fast enough to keep the high-energy-yield aerobic metabolism of the muscle cells in operation. However, the sprinter has kept moving for a time by using the primitive anaerobic system. The muscle cells cannot burn the glucose used for energy all the way to carbon dioxide and water, but accumulate lactic acid, somewhat in the way that the yeast cell accumulates alcohol. The lactic acid serves as a signal to the nervous system that it is time to quit. The overconfident runner rests for a while, and enough oxygen is brought in to burn up the lactic acid, restore his reserves of glycogen (which is made of glucose), and he can be on his way again.

The energy-yielding metabolism of all organisms, from bacteria to man and whales, involves shifting hydrogen from one compound to the next, from a higher energy level to a lower one. In anaerobic life, the bottom is reached with such compounds as alcohol, lactic acid, and other fermentation products. During the course of anaerobic metabolism, 7 ounces (200 g) of glucose will yield about 22,000 calories. In aerobic, or oxygen-using, metabolism the bottom is reached with the combination of hydrogen and oxygen to form water. This extracts the maximum amount of energy, 7 ounces of glucose yielding 267,000 calories.

So complicated are bacteria that it is clearly hopeless to try to assemble them as one would an automobile. What has to be done is to set self-reproducing systems such as the protenoid spheres or coacervate droplets into competition with each other, and let the droplets work it out according to a free enterprise system. At the same time, the chemical environment could be biased toward increasing the chance that selection in favor of nucleotide aggregates would occur. It seems quite possible that bacterialike cells could be produced in this way.

Origin of Modern Life

In recent years, the problem of the evolutionary origin of the eukaryote cell—from the blue-greens, the bacteria, or both—has been a matter of discussion. One theory is that the eukaryote cell evolved by the gradual internal improvement of cell structure. For example, the parts of the DNA loop that had to do with synthesizing photosynthetic pigments, and some of the enzymes associated with the chemical aspects of photosynthesis, pinched off to form plasmids (intracellular particles that can reproduce themselves). These plasmids gradually became the complex chloroplasts of the eukaryote algae. The genes responsible for the set of enzymes involved in the use of oxygen separated off to form the mitochondria. As a modern analogue of such events, one can point to the nitrogen-fixing bacteria, which have genes that code for the enzyme complex that incorporates free nitrogen into organic compounds. Here the DNA segment involved pinches off to form nitrogenase plasmids, which can be transferred in good working order even into other species of bacteria.

On the other side of the controversy are those who believe that the eukaryote cell is an aggregate of different kinds of bacteria and blue-greens that have fused together to form a complex cell surrounded by a single membrane. For example, bacteria that engulf blue-greens, perhaps at first as food, could produce a system in which the two would mutually tolerate each other. These two would then function as symbionts, living together in mutual benefit, the blue-green furnishing energy from sunlight for food production, the bacterium producing supplementary pathways for more diversified food production. The bacterium might also confer upon the guest blue-green the ability to escape from enemies or move into better habitats. Blue-greens are only capable of a slow, gliding movement, whereas many bacteria swim rapidly with flagella that pull them through the water. Anaerobic bacteria might team up with those adapted to use oxygen to produce a combination with a broader nutritional spectrum. If one of the kinds of bacteria and blue-greens in a complex association of three or more types took over the function of serving as headquarters for the DNA coding section, it could gradually evolve into the structure seen as the nucleus of the modern cell.

Probably the symbiont theory of the origin of the nucleated eukaryote cell is on the surface the more attractive of the two theories. Yet when cell biochemistry is examined in detail, the symbiont theory runs into difficulties, since there are many basic differences between the functioning of the cell organelles (cellular structures such as mitochondria and chloroplasts that correspond to the organs of larger animals) and the functioning of primitive prokaryote cells. This may be the result of a long and tortuous evolution by the symbiotic route in which the original partners of the symbiosis have been much changed. On the other hand, it may mean that internal differentiation within the cell, also by a long evolutionary history, has produced organelles that are reasonably good facsimiles of independent bacteria or blue-greens.

The first half of the Precambrian saw the origin of life and the origin of cells with DNA-coded enzymes. It saw the invention of ways to use solar energy for life. Our doubtless very imperfect view of these far-distant times sees the waters of the lands and seas populated with the prokaryote cells—bacteria and the blue-greens.

The second half of the Precambrian saw the evolution of the eukaryotes, living in a world in which the vast stores of nonliving organic compounds were depleted, and in which most were appropriated into the bodies of living creatures. The air was much like that of today, rich in oxygen, which infused the biological world with the restless energy that is the mark of modern life. Nitrogen was in the air as free nitrogen, produced by various kinds of bacteria, without the primitive ammonia (posited by some theorists) which would be poisonous to the oxygen-breathers.

Most important, during the last part of the Precambrian the many-celled plants and animals came into existence. The first many-celled organisms were probably little more than colonies in which cells that differed little or not at all from one another were held together by their sticky cell membranes. In one group of present-day protozoans (single-celled animals) there are colonies of two, four, eight, or many cells. The only obvious advantage of such an arrangement is sheer size. Another microbe might well be able to engulf a single one of its neighbors but would give up a cluster of four stuck together as a bad job. In some of the larger colonies, there is differentiation between cells; those that have longer flagella tend to do most of the work in swimming. Other colonies have certain cells equipped with eye spots that help the colony steer toward the light, and give it a front and back.

Most of the 2 million or so known living species of animals, and another million of plants, are many-celled organisms composed of eukaryote cells that are of many different kinds, specialized for different functions. Whereas the problems that faced the microbes at the molecular or chemical level were simply nutritional, those facing the many-celled organisms are concerned in addition with embryological development of thousands or trillions of cells of a great number of kinds and arranged in definite patterns from a single mother cell, the egg. The nucleus contains an immensely expanded store of DNA tape compared to the prokaryote. This superabundance of DNA probably should be characterized as a vast bureaucracy that makes the decisions as to which of the genes will come into action at what time and what place during development. This has been the main concern of evolution from the late Precambrian until the advent of man. And the nature of this bureaucracy (or clever, simple scheme that operates democratically) remains the deepest mystery of modern biology.

FOUR

·

LIFE PREVALENT

Life in the Sea

As we now understand earth's history, the modern ocean (as a unit, not in details of its shape) formed from a cooling atmosphere early in the third eon B.P. (an eon is 1 billion years, so this would be a little less than 3 billion years ago). Life appeared 100 or 200 million years later, and modern eukaryote organisms appeared at the beginning of the second eon B.P. The seas, perhaps with minor and local exceptions, remained the home of life until about 500 million years ago, when the invasion of the continents by plants and animals began on a large scale. Today the ocean is continuous, occupying mainly the Southern Hemisphere, with the Pacific and Atlantic forming two large embayments.

The ocean forms 70 percent of the surface of the earth and is of immense volume, so that if the solid earth were graded flat, it would be "water, water everywhere, nor any drop to drink." Nevertheless, space available for the primary production of biological materials by plants in the ocean is limited. The volume that is bathed in sunlight is relatively small. Even the superlatively clear waters of the open sea are not transparent. A 3-foot (1 m) layer of seawater filters out a third of the light. At 150 feet (46 m), scarcely 1 percent of the light remains, and this is mainly blue light. Below this depth, the last vestiges of light fade rapidly. Most of the volume of the marine environment is black and icy cold. Only here and there is it lighted by the strange luminous animals that prowl these depths. Near the coastlines turbid river waters and the abundant small animals and plants floating in the water make it relatively opaque, so that the sunlit layer is even thinner. So far as abundance of life is concerned, this disadvantage is more than compensated for by the richness of dissolved mineral nutrients. In shallow seas near the coasts marine life is richer than in the high seas.

Only a minute amount of the sunlight absorbed in the upper layers of the open sea is trapped and used for photosynthesis by plants. The remainder is converted into heat. It is mainly the sunlight absorbed by seawater that keeps the earth warm through the cold seasons, since land reflects most of the light falling on it. The ocean is our solar heat storage system.

Land and sea are alike in that about half their respective areas are biological deserts or semideserts. On the land, there are the vast regions where 15 inches (38 cm) or less of precipitation falls per year. Here, at least in warmer regions where evaporation is high, the vegetation is generally sparse short grass, and in the true deserts that are rainless for months or years at a time plant life may be widely scattered or completely absent for mile after mile.

The open sea—nearly all of the Atlantic and Indian oceans, and the Pacific on either side of a narrow band of luxuriant life along the equator—also is a biological semidesert. Here the amount of life is restricted chiefly by the scarcity of the element phosphorus. Although it is abundant in seawater near the ocean floor and near the coastlines, there usually is no way for phosphorus to reach the surface waters from the great depths of the open seas. These surface layers generally are isolated by a boundary layer of subsurface horizontal currents. In some localities, as in the equatorial Pacific and along the west coast of South America, strong upwellings do penetrate to the surface, bringing with them the life-giving element.

These deserts or semideserts of the open sea also are poor in nitrates. But this is not, in the long-range view, the limiting factor. There are a number of aquatic microbes, most importantly the blue-greens, that have a set of enzymes capable of combining the nitrogen dissolved in the water (from the atmosphere) with other elements to form biologically usable compounds. If phosphorus were abundant in this upper layer of the sea, it seems unlikely that the nitrogen-fixing, photosynthetic blue-greens or similar organisms would not have invaded this environment and diversified on a massive scale, thus dominating the most spacious habitation on earth. However, the DNA mechanism, based on phosphorus, and the high-energy phosphate mechanism involving ATP are firm requirements for life as we know it. Any alternatives that may have existed in the early history of life were presumably replaced by their phosphorus-equipped rivals.

The heaving mats of brown seaweed that occur on many coasts are watery jungles, with plants stretching 50 or 100 feet (15–30 m) above the sea floor. Coral reefs and the cold waters of the North Atlantic fishing banks teem with life. But despite such local richness, there is more living material, by weight, on land than in the sea.

Of an ineffable blue are the sunny meadows of the

high seas. Their plants are invisible, being microscopic and spaced far apart. In the pioneer explorations of marine life during the nineteenth century, these plants were collected by towing cone-shaped nets, made of cloth with very fine mesh, for some miles through the clear water. Such nets gathered an astonishing variety and number of small creatures: a majority consisting of plants, a minority of animals, and a number that were neither or, more accurately, were both plant and animal. Nearly all were of microscopic size. The chances of these relatively small, slow nets catching one of the good-sized animals that roam these waters were rather remote. This assortment of tiny plants and animals is called "plankton," an assemblage of organisms that floats more or less passively in the sea, carried by currents both horizontally and vertically into different sorts of environments in rhythms that may correspond with or be immensely longer than their life cycles.

Usually the most common of these microscopic plants in the net are the diatoms, often called the "grasses of the sea." The single cell that makes up one of these plants is enclosed in a jewel box of clear, hard silica (roughly speaking, glass), with a lid that fits down over and encloses the sides of the base. This box is engraved with hundreds or thousands of lines and dots arranged in beautifully symmetric patterns that are admired by students of these algae and are used to classify them. These minute plants probably can control their buoyancy by regulating the droplets of oil that they secrete into their bodies, but not much is known of the world as experienced by the diatom. In some parts of the ocean, dead bodies of diatoms fall slowly to the sea floor in astronomical numbers. In California, near the town of Lompoc, are beds of soft white rock hundreds of feet thick that are composed of diatom shells. These beds are owned by a corporation that sells the rock for a multitude of uses ranging from abrasives to paints to insulation.

We tend to think of the abundant microbes of the sea, such as the diatoms, as ancient and little-changing. But on the contrary the dominant plants of the marine meadows are, like the grasses of the land, of relatively recent origin. Diatoms did not appear in the geological record until the Triassic, a mere 200 million years ago, and did not become reasonably abundant until the Cretaceous, the period that saw the disappearance of the dinosaurs.

Next most abundant after the diatoms, or in some places the most abundant, are actively swimming microbes called dinoflagellates. They have equipment for photosynthesis and have a simple photosensitive spot that orients them to light. When near the surface they are in the domain of the botanists, for they live like a poet on air and sunshine. When they are carried, or swim, to deeper and darker waters they express their animal side and feed on helpless microplants or on other small animals by engulfing them.

During the twentieth century, explorers of the deep came to realize that their plankton nets were not telling the whole story and that many or most of the plants were too small to be caught. New methods using centrifuges or ultrafiltration disclosed a world of very small plants, as small as a micron (0.000039 inches) in diameter. The smallest of these appear to carry the minimum of equipment for a eukaryote (or nucleate) cell that is able to carry out photosynthesis.

Using the array of equipment needed to get a reasonably complete sample of the plankton, scientists determined that in most of the open ocean the concentration of plants was relatively low just at the surface, rapidly increased to a maximum about 80 feet (25 m) below the surface, then fell away to vanishingly small amounts at 500–650 feet (150–200 m). Numbers of organisms varied from place to place. The main large-scale anomaly was the central band of upwelling in the Pacific, which produced unusually high concentrations of microplants. The seaweeds of the Sargasso appear to be mainly a concentration of coastal seaweeds torn away by storms and gathered in a gigantic eddy in the Atlantic. There is some controversy about the degree to which this population is able to reproduce itself in the open sea.

Generally, the abundance of microorganisms is higher at higher latitudes, and falls off toward the equator. One rough estimate characterized the Atlantic plankton at 20° north latitude in this way: A cube of water in the upper 165 feet (50 m), measuring about 3 inches (8 cm) on a side, averaged about 20 million organisms. The number appears high, but it should be remembered that the organisms are very small. Allowing an average size of 2 microns for an organism, its nearest neighbor would be about 8,000 times its own length away. Translating into human dimensions, everyone would have a nearest neighbor something over a mile away.

Animals that graze such sparse meadows have a problem finding food. There are no large animals, analogous to the sometimes gigantic grazers found on the land, that can feed directly on these microscopic and scattered plants. Most of the marine animals that graze on them belong to four groups, ranging in size from a fraction of a millimeter to a few centimeters. These are the radiolarians, the foraminfera, the larvaceans, and the copepods.

Radiolarians are single-celled organisms, although

sometimes with many nuclei or aggregated into simple colonies. Like many of the simpler animals, they are careless of their individuality; they sometimes allow other and smaller microscopic creatures, especially plants, to dwell inside their living substance, to what is assumed to be the mutual benefit of host and guest. Some are large enough to be seen without a microscope, being flecks of matter 2 or 3 millimeters (about ⅛ inch) in diameter. They are among the most elegant small animals of the sea, resembling the artist's conception of stars, or elaborate three-dimensional snowflakes. Basically, they are spheres surrounded by a multitude of long radiating filaments of protoplasm, so that an enormous surface area is presented to their surroundings as they drift through the water. When smaller creatures are touched by the filaments, they are engulfed and devoured.

Often the protoplasm of the radiolarian is supported by an intricate internal skeleton of silica. As with the diatoms, these hard, resistant skeletons accumulate in some regions of the ocean in countless numbers on the sea floor, where they make a radiolarian ooze. This may be brought to the surface by geological processes, to form a radiolarian earth, analogous to diatomaceous earth, or may be transformed into nodules of flint, where the skeletons may still be observed. Radiolarian skeletons have been found in rocks of Paleozoic age.

These animate, three-dimensional plankton sieves were first adequately described as living animals by T. H. Huxley (1825–1895) in the mid nineteenth century. Their chief admirer and explorer was another contemporary disciple of Darwin, Ernst Haeckel (1834–1919), a German who sought temporary refuge from his continuing political and ideological battles with the authorities to study these beautiful creatures. He described 4,000 new species of radiolarians (a name he invented) and celebrated them in two huge illustrated volumes.

Foraminifera, or forams, are single-celled animals that average about a millimeter ($1/25$ inch) across but they construct a vaguely mollusclike shell that may attain a few centimeters (an inch or two) in length or diameter. Nearly all the species are bottom dwellers, but the minority that float in the upper layers, buoyed by oil droplets or bubbles, occur in large numbers. Nearly 30 percent of the ocean floor is covered with a layer rich in their shells, called Globigerina ooze and named for a common foram of the upper layers of the sea. The shells are abundant fossils, some extending back to Cambrian times. Their close study and identification are essential parts of the science of stratigraphy. On Mount Everest, at 22,000 feet (6,700 m), is a layer 200 feet (60 m) thick of fossil foram shells of Eocene age. Eocene forams called nummulites make up the limestone blocks of which the great pyramids of Egypt were built.

Forams have bristling sunbursts of extended protoplasm, resembling those of the radiolarians, and like them trap microrganisms as their bearers drift passively through the sea.

Our third group, the larvaceans, are small, active, muscular, many-celled animals only a few millimeters long. They are near the most distant ancestry of man, having what might be called a boneless backbone, which gives support for the muscles that move the minute but powerful swimming tail. This supporting rod or notochord gains its strength from cells stretched drum-tight by water that is drawn into and kept in the cell by specialized physico-chemical processes. It was some time before biologists realized that the animal itself was the manufacturer, engine, and pilot of a large sieving apparatus that enables it to eat its way through the sparsely inhabited universe of the open sea. The larvacean constructs a streamlined, movable house, resembling a fat submarine, of a gelatinous substance that is mainly water. So economical is its construction that the animal can afford to shed it and construct another every few hours. Without careful observational techniques, this superstructure collapses into an indecipherable glob when the animal is taken out of the water. In it are fine sieves, an external coarser one that filters out large particles, and an inner fine sieve with a mesh size of about 1 micron that can filter out micro-microplankton. This underwater houseboat is furnished with a trapdoor which holds the inflowing water brought in by the swimming tail until a certain amount of pressure is built up, then suddenly releases, driving the structure forward. Some of the very small microbes that make up a large part of the plankton were in fact first seen alive in the sieves of larvaceans, since they had passed through the much coarser mesh of the plankton nets of the early explorers.

If a net, similar to a plankton net, is swept through a grassy meadow on land, it collects a swarm of small, buzzing insects. The marine plankton net can be towed for weeks through the waters of the high seas without catching a single insect. In the open seas there are almost no insects—only a group of small water striders that live on the surface. Their place is taken by a rather similar group of animals, the copepods, whose nearest relatives are the crabs and shrimps; all these are members of a group called crustaceans. Copepods include 2,000 to 3,000 species in the waters of the high seas, and another 8,000 in coastal or fresh waters and other more specialized habitats. Insects, by contrast, have about 800,000 described species, and probably 5 to 10 times this number remain undiscovered. This gives a fair idea of the difference in diversity of habitats on land and in the sea. Although the sea is far less uniform than is generally believed, there is a much greater variety of habitats, often small and well-delimited, on the land.

Copepods are generally a few millimeters long, and those that feed on plankton graze by beating a set of microscopic combs several hundred times a minute in such a way as to drive the microplankton into their mouths.

In Arctic and Antarctic waters there are immense numbers of larger plankton-grazing crustaceans called euphausids. Huge heaps of these animals are tumbled out of the stomachs of slaughtered baleen whales.

There is little doubt that the vast stretches of the Pacific have little to offer man. During a 20-day voyage across the Pacific on a high-decked ship, the only animals I saw were the beautiful blue flying fish, which rose up from the water like the grasshoppers of a western prairie. Nor did we see a fishing boat (or any other kind of boat) or an airplane. A glimpse of a tiny island and a single seabird only reinforced the loneliness. Man's chief activity in these waters now seems to be to infest them with plutonium-carrying submarines and to sow them with sensing devices that locate these mechanical monsters, friend or foe.

However, the scanty pastures of the open sea, like the short-grass prairies of the American West, support a diverse population of animals, large and small. On the land, insects and mice graze the vegetation. Other insects, birds, foxes, coyotes, and wolves prey on these creatures. There also are large herbivores, such as buffalo and antelope, that feed directly on the vegetation, and these used to be the prey of the relatively rare and large carnivores:

the grizzly bear, once king of the western plains, the mountain lion or panther, and the wolf.

It is just so in the sea, although the minutely divided plant life in the open ocean cannot be eaten directly by large herbivores. Man himself has been unable to devise mechanisms efficient enough to make it pay to filter out the microbial plants. This is done by the small radiolarians, larvaceans, and crustaceans already described, as well as some other animals such as the arrow-worms and the free-swimming snails or sea butterflies, and a few of the smaller fishes. All the larger animals are carnivores, feeding on the small animals or tearing chunks off one another. It is these that provide food for human beings.

The blue whale, largest of all animals living on earth, perhaps attaining a weight of 400,000 pounds, feeds on very small animals. So far as is known, only in polar waters that are rich in nutrients and support a rich soup of plankton can the blue whale sieve out crustaceans fast enough to gain weight. There it gorges itself for a few months during the short summer. Then it heads for tropical waters of low latitudes, where it bears its young and apparently does not feed for as long as 9 months.

Ahab sought his enemy, Moby Dick, along "the Line," in the rich equatorial upwelling of the Pacific. The sperm whale is a carnivore that feeds mainly on squid, which are powerful, predatory molluscs, often of gigantic size, the muscular tentacles and body sometimes nearly as long as the whale itself. These large squid may hunt hundreds of feet below the surface, and the whales dive after them. Below the lighted zone the whale must locate them by sonar, and the death struggle takes place in frigid blackness.

There are several steps in the chain of predators, from smallest to largest, in the waters of the high seas, where there are no herbivores larger than some of the herrings. The adaptations for capture and escape are marvelously varied and ingenious. In the swift tuna fishes the steering fins, when not in use, fit into slots like the retractile landing gear of an airplane, and the eyes are flush with the curve of the skin, to complete the streamlining. Tuna swim with astonishing speed, not by violently thrashing the body, but with a trembling vibration of the hard and elegantly designed tail fin, which is driven by tremendous banks of muscle that give off enough heat in their ceaseless operation to make this fish a warm-blooded animal.

The feeble jellyfishes, 99 percent water, swim slowly by graceful pulsations that produce a smooth succession of jets. But some are formidable predators. One blue and orange species has a cartwheel body 12 feet (3.7 m) in diameter and tentacles 100 feet (30 m) long that are armed with millions of venomous microscopic barbs that can kill a man.

To avoid the daily carnage as much as possible, many of the medium-sized predators lurk in the darker lower regions during daylight hours, then come up at night to feed on their smaller and less mobile victims. However, they are to some extent thwarted in this stratagem by the fact that the millions upon millions of microbial plankton are often luminescent, bathing the sea in a blue or golden light. It has been suggested that the ubiquity of luminescence in the microplankton is an adaptation that lights up the scene so that the enemies of the grazing animals can catch their prey all the more easily.

At death the sea creatures, great and small, finally settle toward the bottom. Enough reach their destination to create a zone of nutrient enrichment at the sea floor that supports a population of thousands of species of animals, many of them primitive, and many of them bizarre.

Most are small, but the carcasses of large animals from above attract scavengers of respectable size. A most striking fact concerning the animals that live on the deep sea floor is that most species are luminescent. Some have patterns of lights that seem to be used for species recognition. Some puff out a cloud of luminescence that momentarily confuses a pursuer. Some have searchlights, with lenses, used to help find prey. Others have lighted lures to attract their victims.

Deep-sea fishes that in drawings look like dragons may be feeble, soft creatures a few inches long. They are equipped, however, with wide gaping jaws set with long, needle-sharp recurved teeth that cannot let go of anything that the jaws have touched. The fish behaves as a thin-walled distensible bag that is pulled over its prey, perhaps even larger than itself, by the alternate forward movement of the jaws. These reset the teeth in such a way as to force the victim inexorably into the fatal hydrochloric acid bath of the stomach.

The larger predators of the open sea have been mainly vertebrates—fishes and mammals and, for a time, reptiles. There is some question about where the fishes originated. Probably it was the environmental challenge presented by the strong currents of rivers emptying into the sea that converted the boneless marine prevertebrates into vertebrates. These then reentered the oceans and ruled for hundreds of millions of years before the marine reptiles appeared.

The second wave of "rulers of the sea" was reptiles (some as large as the largest fishes) that invaded the oceans during Mesozoic times. Some imitated the form and life style of large fishes. Others were the venerable turtles, which invaded the seas in the Cretaceous. Plesiosaurs had a flat body, swimming legs, and a long neck, useful in the final capture of prey. These were as much as 50 feet (15 m) long. Finally, near the end of the Age of Reptiles, close relatives of modern lizards invaded the sea, and quickly evolved into giants some 30 feet (9 m) long. Except for the large sea turtles, the last of the giant sea reptiles abruptly disappeared at the close of the Mesozoic.

At that time the continents had moved northward and southward and swung away from one another into approximately their present positions. This change coincided with the extinction of the dinosaurs on land and the large reptiles in the sea. The reasons for these extinctions are not known, but perhaps the partial isolation of the north polar sea from the rest of the ocean, and the replacement of the south polar sea by the continent of Antarctica, would eliminate the heat-storage capacity of the huge volumes of water represented by the two polar seas and would drastically alter the climatic regime of the world.

With the sea cleared of the giant reptiles, the way was opened for invasion by the mammals. A variety of types evolved. Such forms as the polar bear and sea otter are only slightly altered for life in the sea, and intermediate types such as seals can live far out at sea indefinitely, but must come out on land to breed and raise their young.

Whales and their smaller relatives, the porpoises and dolphins, began their evolutionary history in the Eocene, some 50 or 60 million years ago. Whales were at first rather awkward, serpent-shaped animals, but the modern whales, the largest of all animals known, have the blunt-faced contours found in the fast nuclear-powered submarines. These astonishing creatures roam the oceans at will and communicate over distances of hundreds of miles by sound. They sing complex songs that may last for nearly half an hour. Huge-brained, they are in their way perhaps as intelligent as ourselves, but unlike us, who are destroying them, they are not cursed with the fatal sin of unlimited greed.

Life on the Land

Because plants in the sea are of microscopic size except in waters along coastlines and in such exceptional areas as the Sargasso Sea, it is the animal life that in appearance dominates the marine environment. Quite the opposite is true of the land.

On land, usually the plants dominate, a green matrix in which is embedded more or less unobtrusively the animal life that depends on green plants for its sustenance. Perhaps a third of the surface of the earth is (or was, in the natural state) covered with forest, ranging from the gloomy covering of spruces and firs of the far north to the immense tangles of tropical jungles, from the shady stands of temperate deciduous forests that once covered nearly all of the United States east of the Mississippi to the chaparral of junipers and cedars that are scattered over the sand and rock of the arid American southwest. In the tall forests, the upper layers of this three-dimensional environment are still poorly known and probably never will be known, since they are being destroyed faster than the growth of scientific natural history.

An Oriental version of the Biblical verse

Consider the lilies of the field, how they grow;
They toil not, neither do they spin.

is Lao-tze's observation of the sixth century B.C.: "All things in nature work silently. They come into being and possess nothing. They fulfill their function and make no claim. All things alike do their work, and then we see them subside. When they have reached their bloom each returns to its origin. Returning to their origin means rest, or fulfillment of destiny."

Carolus Linnaeus (1707–1778), the scientific Adam who named the kinds of living things, says of the differences between plants and animals:

VEGETABLES clothe the surface with verdure, imbibe nourishment through bibulous roots, breathe by quivering leaves, celebrate their nuptials in a genial metamorphosis, and continue their kind by the dispersion of seed within prescribed limits. They are bodies organized, and have life and not sensation.

ANIMALS adorn the exterior parts of the earth, respire, and generate eggs; are impelled to action by hunger, congeneric affections, and pain; and by preying on other animals and vegetables, restrain within proper bounds the numbers of both. They are bodies organized, and have life, sensation, and the power of locomotion.

A textbook on earth's history is likely to say that the Precambrian landscape was barren of plants. The oldest known fossils of land plants are of Upper Silurian and Lower Devonian age, roughly 400 million years old. These were tough-stemmed flowerless plants 3 feet (1 m) high. But it seems reasonable that the land surface during the late Precambrian—rock, sand, and even a primitive kind of soil—had some cover of ground-hugging plants, perhaps no more than layers of colonial single-celled microbes that provided great expanses of greensward (in the original sense of sward, "skin of the earth"). The burning ultraviolet radiation of the early Precambrian presumably was softened toward the end of the era, as photosynthetic marine plants began to produce large quantities of oxygen, which screens out the more deadly wavelengths.

Damp or swampy areas surely were green long before plants that would normally be preserved as recognizable fossils came upon the scene. Today, much of the ground

surface of even rather dry grasslands is covered with lichens, which are organized colonies of two kinds of microbes living together—algae and fungi. They have a Precambrian look about them, and from the fossil record it
would be difficult to prove that lichens as well as many
other kinds of simple, low plants did not exist in these ancient times.

The environment of the open sea is three-dimensional. By contrast, the biological environment on land
was skin-flat until the plants learned to lift themselves
into the air. Probably in Cambrian times this had not been
accomplished. Three hundred million years later, plants
had lifted themselves as high as 3 feet (1 m). Then,
during Devonian times, the evolutionary race for size,
for escape from the shade of one's neighbors, accelerated mightily, and true forests developed. Within less
than 25 million years, there were vast forests of trees more
than 100 feet (30 m) high, producing such thick layers of
coal that this geologic period is called the Carboniferous.

It would be well to preface a discussion of land plants
with an account of groups of larger marine plants that
were not included in the description of the microbial
plants of the open ocean. These are the seaweeds, which
grow rooted on the sea floor along the coastlines of continents and islands. They are composed of millions of cells
of different kinds, with differentiation of labor, and coop-
eration to produce organisms that may reach several
hundred feet in length. The diver at the floor of one of
these seaweed jungles sees slender, curving trunks that
rise to leafy tops scores of feet above his head. The seaweeds, like their pigmented, photosynthetic, single-celled
relatives in the seas and lakes, are termed algae. This
term has its historical origins in the need to distinguish
them from the fungi, which are simple, eukaryote plants,

often with colorless cells, that are not photosynthetic and have to feed on other organisms or their decay products just as do the animals.

Seaweeds are an abundant and diversified group. Along the Monterey coast of California 400 species have been identified, perhaps half of the number of land-plant species in a comparable area inland from the coast. There are three main groups of seaweeds, classified simply as the reds, browns, and greens. The differences in pigmentation are related to many biological features that are characteristic of each. Green seaweeds are a minor group, usually small and uncommon plants, but with some abundant forms such as the sea-lettuce of shallow northern waters. The many species of red seaweeds also are generally small and not abundant. The bulk of the forests of the sea is made up of the brown algae, which include the giant kelps. Washed up on the beach by a storm, they collapse into tangled masses. They are held up like trees in their natural habitat by small balloon-like floats at the tips of stems. These are so airtight that they pop when stepped on. Seaweeds do not have true roots, in the sense of roots of land plants that absorb nutrients and water from the soil. Seaweed roots are only mechanical devices that anchor the plant. The nutrients necessary for photosynthesis and life—carbon dioxide, water, minerals, and oxygen—are absorbed directly by leaves and stems from the surrounding seawater.

In most single-celled plants and animals there is much reproduction, but little of it is sexual. Usually reproduction involves only cell division, or fission. This results in offspring that are genetically identical with each other and with the parent. Exceptions arise by mutation, which is a defect in the duplication of the DNA tape or tapes (the chromosomes, in a eukaryote cell) that occurs about once in a million replications.

Sexual reproduction, in which two cells or their DNA components fuse before reproducing by fission, usually guarantees that the offspring will be of different genetic makeup. This variability comes about because the genetic material of each of the parents, except under unusually incestuous conditions, has a different historical background and may differ in hundreds or thousands of genes that have accumulated by indefinitely long periods of mutation. As a result of certain mechanisms of chromosomal behavior, in a single population of a species there may be billions of individuals among which no two are exactly alike. Many of the microorganisms and smaller many-celled organisms may reproduce without sex for many generations, then stir up the genetic mix with a round of sexual reproduction. Reproduction in the large animals, which have small populations of a given species as compared with the microorganisms, is almost invariably sexual, although in theory it should be possible to produce embryos of any organism by artificial means without fertilizing the egg.

There is an obvious flaw in sexual reproduction: if the parents had good enough genotypes to survive to maturity, why destroy them by sexual mixing? This paradox is the key to an understanding of historical, that is, evolutionary, biology. There is a saying among biologists that biology produces highly exact or perfect results by highly inexact or imperfect means. In the world of living nature, many individuals find themselves betrayed by their genotype, and die young. But a few miles, or a few years, away they might have found themselves in a situation in which their genotype made them winners. Often, premature death appears to the casual observer to be only a matter of bad luck, but careful analysis generally shows that this natural selection is a nonrandom process, that the results are biased as a result of characteristics of the environment. Sex is an adaptation by which a species throws essentially infinite variety of structure and physiology at the environment, with the result that environmental differences in both time and space—and they are an outstanding characteristic of the surface of the earth—are met successfully by enough individuals that the species persists. When this does not happen, the species becomes extinct. Survival in a varied and changing world comes only with the premature death of countless individuals. Thus, failure is an absolute correlative of success in our imperfect and changeable world.

Since plants such as the seaweeds and the rooted land plants can not move around, sex is a problem. Some of the seaweeds that live in dense colonies reproduce in the same manner as a swarm of oceanic fishes. They simply release swimming eggs and sperm in huge quantities, these swim about in the water until they meet a partner, and the fertilized egg sinks to the sea floor to grow into another seaweed. Often, however, perhaps as an adaptation to overcome greater distances between prospective parents, the seaweed performs the maneuver in two stages. In the first stage, the plant produces spores that swim or are widely dispersed by currents. These settle to the bottom, where they germinate and produce smaller plants that do not produce sheer bulk of plant matter, but are specialized to produce sperms and eggs. These small plants can be much more numerous than the parents in a given habitat or area, and therefore sperms and eggs from different parents have a better chance of meeting than if they had been produced by huge, more solitary parents.

Lack of mobility in seaweeds also affects the problem of dispersal, of making it possible for the young to find living space. The seaweeds accomplish this primarily by producing spores—small, long-lived cells that swim and drift in the currents, sometimes for great distances.

We can now leave the plants of the sea, and consider the history of the land plants. It is generally believed that the multicellular land plants evolved from one or several groups of seaweeds, but this is by no means certain. Such plants conceivably could have arisen in fresh water or damp places. Such small plants as the mosses or liverworts could have gradually evolved from the single-celled or colonial algae that probably lived on the continents for hundreds of millions of years before the oldest land plants actually known in the fossil record appeared.

All life is based on carbon compounds, and this carbon is ultimately obtained from one of the least common gases in the atmosphere, carbon dioxide. Air is composed primarily of four gases. These are nitrogen (77 percent), oxygen (21 percent), argon (0.8 percent) and carbon dioxide (0.03 percent). Carbon dioxide is highly soluble in water, so the oceans contain a reservoir of carbon dioxide greater than that existing in the atmosphere. It seems incredible that a component amounting to only $3/100$ of 1 percent of the atmosphere filters through the leaf surfaces fast enough to produce the great woody trunks of forest trees. In phytotrons, specially designed rooms where all environmental conditions can be controlled, plant growth can be stimulated considerably by increasing the carbon dioxide concentration. It is possible that the carbon dioxide concentration in the atmosphere has varied considerably in the geologic past.

Although animals and microbes in general balance out carbon dioxide loss to plants by eating plant material and converting it back to carbon dioxide as they burn it metabolically, there is a slight net loss. This is caused by burial of organic materials under muds and sands, beyond the reach of decay bacteria. Theoretically, the oxygen in the atmosphere is sufficient to burn all the oil and coal in existence and thereby to create great amounts of carbon dioxide. However, most of this fossil carbon is too finely dispersed through the rocks to be recovered for fuel. The oil shales appear to represent a marginal situation. Volcanic activity, by destroying limestone, may return to the atmosphere some of the carbon dioxide lost to limestone formation, and may even provide additional amounts from deep carbon-containing rocks. An immense amount of carbon is stored in the wood of trees.

One of the main challenges facing the early land plants was how to reach up for light, but the solution of this problem made sexual reproduction more difficult. In low, ground-hugging plants such as mosses, sperms and eggs are produced. Eggs are heavy and sedentary, the sperms light and motile. They move by lashing their long tails and swimming through water. A rainstorm provides enough water for the sperm cells of mosses to get moving. But by the time plants evolved to a few or many feet in height, as in some of the ferns, a Noachian flood would have been required. The ferns solve the problem by producing spores by the millions, diminutive cells with a tough coat and persistent spark of life, so light that they can be carried hundreds of yards or even thousands of miles in air currents. Some alight in damp places favorable for germination and growth. They produce, not another fern, but a small flat disc of a plant, sometimes green and living in the sunlight, sometimes underground and living like a fungus on decaying material. The business of these plants is the production of eggs and sperms,

and here the rains produce the conditions in which ga-
metes can function. The product of fertilization grows into
a fern, and the cycle is repeated. Those two forms are the
sporophyte and gametophyte generations.

Living plants exhibit enough variations on the theme
of alternation of sporophyte and gametophyte generations
that several different historical pathways to the modern
plants can be suggested. In principle, they involve reduction in the size of plants in the gametophyte generation to
the point where the male, sperm-producing plant becomes
what appears to be a single cell. The minute male plant is
the pollen grain. In the most primitive pollen-bearing
plants, it is supposed, the grains were produced in great
quantities and were carried by the wind. The grain perhaps alights, with one chance in millions, on a receptive
female plant that is also small and cupped in the center of
a flower.

In the seed plants the tiny female plant that lies in
the flower becomes the center for the growth of material
that surrounds, protects, and nourishes the embryo.
Flowers of pines (perhaps technically not true flowers)
and grasses are scarcely recognizable as flowers in the
common sense of the word, but those of most species of
other land plants are large and generally pleasing to the
eye, with graceful geometric patterns, elegant colors, and
often pleasant scents. Such flowering plants became common during the Cretaceous, a time when the continents
were separating from an original protocontinent of the late
Paleozoic and drifting toward their present positions. The
ancient fern and cycad forests were mostly exterminated,
with only remnants left in the more southern latitudes.

At the same time a group of insects called the Hy-
menoptera, a group that includes the bees and wasps, also
picked up an active role in the drama of biological evolu-

tion. There now exist some 20,000 species of wild bees. Most of these feed their helpless young a mixture of honey and pollen, which they gather from flowers. Bees are highly intelligent insects that can learn the geography of a considerable area, which they search for flowers, some specializing in only one kind of flower, others using whatever is available. They may make scores or hundreds of trips between the flower fields and the skillfully constructed nests which they have built for the young, and visit innumerable flowers to gather the food store. Although many other groups of insects and other animals also are involved in transporting pollen, it is mainly the bees that have made a precision job of pollen delivery, taking over from the vagrant breezes.

Land plants directly provide the food base for a host of animals. They are eaten alive by small herbivores, such as insects or mice, on up through a range of larger herbivores ending with the elephants. One African told me of watching elephants apparently resting, leaning against good-sized trees. But suddenly with a snap the trunk of a tree would break and crash to the ground, where the elephant could grind up whole leafy branches with its powerful molars. The elephant had been hard at work.

Except where there are local epidemics of insects, the forests are still green at the end of the growing season, and most leaves die and fall to the ground unscathed. Obviously the primary herbivores, which attack living plants, do not use all the wealth produced each year by photosynthesis. There is no precisely defined balance between herbivores and plants. No two years are alike; one year will be favorable for a given species of insect herbivore, another a bad one. There may be rough cycles of abundance and scarcity, but they are never exact. Plants also respond to environmental vicissitudes with yearly variations in amounts of surplus vegetation produced.

In the long run, there is a balance between production of organic material by the land plants and its metabolic destruction which returns carbon dioxide to the atmosphere to be used again in another round of photosynthesis. This balance is achieved mainly in the soil, a complex mixture of minerals, organic compounds, single-celled microorganisms, and small plants and animals. With their short life cycles, which may be only hours long, and their ability to remain dormant over long periods, the small creatures are relatively immune to the months-long climatic deviations that have disastrous or beneficial effects on the large animals and on plants with life cycles a year or several years long. When conditions are momentarily favorable, these small creatures immediately come into action with metabolic rates hundreds of times higher than those of the large organisms. When conditions are unfavorable, they can lie dormant, nursing a spark of life until the next earthworm dies, or a rain shower soaks a dead leaf that can provide new lives for bacteria by the millions. It is this huge reservoir of microlife that produces the so-called ecological balance in nature. The interactions of the large and conspicuous animals and plants that we see before us are clumsy indeed in maintaining this balance.

At the end of the Precambrian era the eukaryote microorganisms had probably achieved the full range of sexual manipulations of the DNA code tapes (including the equal division of the apparatus into halves in the process of reduction division, and the restoration of the double complement of tapes in the fusion of sexual reproduction) that are found in modern algae. The basis for the evolution of the control mechanisms used in embryonic development had been laid. There were doubtless, in the late Precambrian, a number of simple many-celled green plants in damp places on the land. These were presumably of a rarity, softness, and distribution that their fossil remains are few. The oldest known many-celled animals (metazoans) are the recently discovered Ediacaran fauna of Australia, 680 million years old. These are less modern in aspect, sometimes to the point of being unclassifiable, than the fossils of Cambrian age long known to paleontologists. However, they are not necessarily considered Precambrian. Some writers simply shift back the boundaries of the Cambrian about 100 million years to accommodate them. The first probable many-celled plant fossils predate them slightly.

At the Precambrian–Cambrian boundary the evolution of the many-celled animals is believed to have proceeded with explosive strength. The most popular theory to account for this is that the oxygen concentration of the atmosphere had reached a critical level, perhaps 10 percent, at which oxygen could penetrate the inner cells of a many-celled creature fast enough to maintain a reasonable rate of metabolism. The technical problems to be overcome still are formidable with high oxygen levels in the air. Even in modern metazoans the metabolic rate falls off rapidly with increasing size because of the difficulty of getting oxygen into the deep interior of the animal.

Probably several groups of invertebrate animals began to invade the land early in the Cambrian, but the invasion did not spread far enough from the coasts or become overwhelming enough to produce a fossil record until Silurian times, nearly 300 million years later. The oldest known land animals are members of the immensely varied phylum Arthropoda. They were already abundant in the Cambrian sea, and in modern times are extraordinarily abundant everywhere, as crustaceans, insects, spiders, and their numerous relatives. The early pioneers on the land included scorpions and insects.

By the end of the Paleozoic the insects had assumed their position as the dominant invertebrate animals on land. Judging from the fossils, most of the Paleozoic insects were cockroaches. Toward the end of the Paleozoic the largest insects ever known were the masters of the air—dragonflies with a wingspan of nearly 3 feet (1 m). One famous student of fossil insects happened also to be a devotee of spiritualism. When on a visit to Harvard, where a well-known example of this fossil dragonfly was in the collection, he asked the curator if he could borrow the specimen to take to a medium who, he thought, could bring the insect to life. But the curator refused, saying, "The damned thing might fly out the window."

Today the birds (which evolved in the Mesozoic) make impossible the existence of such large flying insects, because of their superior strength and speed.

On land today most of the small herbivores (say up to about 2 inches, or 5 cm), are insects whereas the vast majority of the larger ones are mammals. By contrast, in the late Paleozoic most of the large herbivores were reptiles. In the Permian there were vast herds of ox-sized reptiles grazing on the primitive land plants. When the vegetation began to change to seed plants during the Mesozoic, the primitive reptilian grazers gave way to the dinosaurs and their relatives. At the same time, the mammals were beginning to evolve from reptiles. Although many of the larger reptiles became quite mammallike even as early as the Paleozoic, apparently the first true mammals were small and evolved in the Jurassic. They are exceedingly rare as fossils. Mammals of the Cretaceous, at first thought rare, have turned out to be common in certain localities. They were small and apparently swarmed in the meadows like mice today. Most were herbivores, feeding in the shadow of the great dinosaurs that ruled the land.

The end of the Cretaceous signaled the end of the dominance of reptiles and a rapid expansion of herbivorous mammals ranging from large to small; the modern ecological picture on the land was established early in the Cenozoic, with insects and mammals pretty much sharing the vegetation between them. During the Cenozoic, the insects remained fairly static. Mammals changed dramatically, however, and near the end of the Cenozoic (about a quarter of a million years ago) ushered in a new phase of earth's history by evolving the species *Homo sapiens.*

The Human World

Living *Homo sapiens* is not closely related to any other living animal, the existing species of great apes having fossil lineages that remain separate from man's for perhaps 5 million years into the past. In the fossil record our species grades rather gradually into an earlier species, *Homo erectus,* and through that is fairly well integrated with the group of mammals called primates by way of a number of types that existed in early Pleistocene and Pliocene times.

Individuals of an abundant, widespread species such as ours generally show a wide range of differences in structure and color from one another. Some of the variation is random. Aside from sex and age-group categories, there may be orderly variation that expresses itself geographically. From one region to the next, individuals of a species of mammal may be on the average larger, or darker, or have longer hair. The museum specialist can sometimes make a good guess as to where a specimen came from just by looking at it.

This kind of geographical variation is analogous to the dialects of a language. The local variants of color or structure are called races or subspecies, and like dialects, have a more or less unique historical background and are maintained by a degree of isolation.

As the human population spread from Africa over the habitable earth, probably reaching deep into South America 10,000 to 20,000 years ago, it became ever more diversified in structure and pigmentation, producing a number of geographic types and subtypes in semi-isolated regions. However, the pattern constantly shifted and the populations continued to mingle, so that irreversible reproductive isolation did not occur. Thus, the human species has retained its cohesiveness as a single species.

The behavioral pattern of an animal results from the continuation of the developmental processes that produced its physiology and structure, except that the external environment now provides a larger percentage of the controlling factors than do the internal stimuli that function during embryonic development. In small, short-lived animals such as insects, the behavioral pattern, although requiring environmental stimuli for its successful development, is to a large extent imprinted on the nervous system. That is, it is largely instinctive. There is a degree of learning, but the insect does not have time for much learning by trial and error and does not depart much from the ways of its ancestors. A widespread species of insect, for example, may have geographic races that differ in color, pigmentation, and behavior, all determined in large part by the differing genetic combinations that evolve in semi-isolated segments of the population.

LIFE PREVALENT • 108

In the evolution of mammals, learning plays an ever larger role in the development of behavior, and reaches a peak in human beings. The resulting cultural differences in mankind are immense. There are differences that result from class structure, differences based on occupation, and on the national milieu, the nation or tribe being a cultural analogue of the race or subspecies in other animals.

A sense of proportion would make it necessary to allocate to the human world, with its appurtenances of countless tools, ornaments, weapons, vehicles, shelters, musical instruments, books and other mental fancies, a diversity in the same league as that of the rest of the living world. The human world, produced by the efforts of mankind itself rather than by natural selection, is in all its diversity the equivalent of millions of species of the world of plants and animals.

When the human world came into existence (we may say, more or less arbitrarily) with the origin of the species *Homo sapiens*, it was more like the restricted universe of a nonhuman species than that of twentieth-century civilization. Like most other species of animals, the human species came into existence gradually. In the evolution of man, the existence of *H. erectus* could be visualized as a plateau, extending from about 2 million years ago up to the late Pleistocene, a third or a quarter of a million years ago. Then evolutionary change accelerated, mostly increasing the brain size and taking the genus *Homo* up a comparatively steep slope to a higher plateau, that of *H. sapiens*. The disappearance of the older species would result from both the evolution of the older species into the next and extinction of the laggards by competition with the evolutionary newcomers.

It is generally said that brain size or quality has not increased during about the past 200,000 years. Human beings with languages of some thousands of words and complex grammatical structure have existed much longer than civilization, which got under way about 10,000 years ago.

This is not the place to discuss the relationship between brain size and intelligence in modern man. Suffice it to say that persons responsible for evaluating the promise of, for example, entering college students do not measure skull capacity.

Homo sapiens was a creditable addition to the fauna of the earth. A powerful, agile, well-shaped mammal, a male human being can outrun over long distances (and especially if he brings along a little food and water) any other mammal. And the graceful porters that carry heavy loads up to the high peaks of the Himalayas are women.

Mankind could be called "singing animals." People spend more time at it, and produce patterns thousands of times more complex and communicative, than any bird.

They could be called "dancing animals," excelling all others in a combination of athletic grace and music.

Sexual activity develops to an unprecedented degree. Human beings spend more time engaging in sex than other animals, and secondary sexual ornamentation and related artistic activity are essentially infinite. In his remarkable novel on the origin of modern man, *The Inheritors* (1962), William Golding writes, "There was no animal on the mountain or the plain, no lithe and able creature of the bushes or forest that had the subtlety and imagination to invent games like these, nor the leisure and incessant wakefulness to play them."

Above all, human beings are talkative animals, babbling by the hour, or listening to highly crafted legends, often shaped into poetry, that tell the adventures of men, women, and deities.

These nonphysical attributes that distinguish mankind are made possible by the enormously developed
brain. Those who study the problem of staying warm at
high altitudes point out that 30 percent of the heat loss is
from the head. This indicates a tremendous flow of ox-
ygen-rich blood through the brain. Perhaps one of the
most serious failings of contemporary industrial civilization is that the brain goes largely unused. Probably precivilized man, taken by and large, came closer to using his
head to its full potential.

The brain, of course, is not the mind. It is a computer
attached to an immense input apparatus. This input
comes from the innumerable sensory nerves that pick up
impulses from the external and internal environments,
and from the chemical-producing glands (roughly, the en-
docrine system) that influence the nervous system. The
brain and the input system together produce the mind.
Mathematical reasoning would seem to be an abstract
operation involving only the brain, but even here there are
phantoms of muscular movements, of balancing, and of
the harmonies of music that arise from incipient function
of the labyrinth of the inner ear.

It is a strange fact that an animal with the seemingly
innocuous characteristics of *Homo sapiens* has produced
a revolution of geological magnitude. Over vast areas the
structure and biology of the surface of the earth have
changed with unprecedented swiftness and magnitude.
For 200,000 years the human species, although truly a remarkable one, lived hidden in the tangled web of terrestrial life. It made no great changes in a world which was
already in the throes of a great geological revolution, the
Ice Age or Pleistocene, which for 3 million years had been
smashing the established order of life and climate. Then,
within the last 100 years, mankind has attacked the integ-

rity of Earth with a ferocity that threatens to reach geological intensity.

Once such cultural objects as ornaments, utensils, or weapons come onto the scene, they appear to take on a life of their own. A flint spear point does not exist merely in the mind of its creator. As a real physical object, it can induce other people to learn to make more spear points.

An exceedingly lifelike property of cultural artifacts is that the archeologist can arrange them in branching evolutionary patterns that develop in time, just as the paleontologist arranges fossil organisms and their living descendents in so-called phylogenetic trees. The human being adds innovations to the next generation of his crafted products; genetic variation and natural selection add innovation or remodeling to already existing organisms to produce organic evolution.

An old theme of fiction is robots built so skillfully that they are able to influence human beings, like Dr. Koppelius's doll. In truth, artifacts much simpler than imitations of life influence their creators. Some naïve visitor from another galaxy might conclude that automobiles are clever organisms that force human beings to both pilot them and produce more automobiles.

The nineteenth-century writer Samuel Butler (1835–1902) develops the thesis of the primacy of artifacts in his novel of a utopia, *Erewhon*. In the chapter called "Book of Machines" he writes:

They [the machines] have preyed upon man's groveling preference for his material over his spiritual interests, and have betrayed him into supplying that element of struggle and warfare without which no race can advance. The lower animals progress because they struggle with one another; the weaker die, the stronger breed and transmit their strength. The machines being of themselves unable to struggle, have got man to do their struggling for them: as long as he fulfills this function duly, all goes well with him—at least he thinks so; but the moment he fails to do his best for the advancement of the machinery by encouraging the good and destroying the bad, he is left behind in the race of competition; and this means that he will be made uncomfortable in a variety of ways.

So that even now the machines will only serve on condition of being served, and that too upon their own terms; the moment their terms are not complied with, they jib, and either smash both themselves and all whom they can reach, or turn churlish and refuse to work at all. How many men at this hour are living in a state of bondage to the machines? How many spend their whole lives, from the cradle to the grave, in tending them by night and day? Is it not plain that the machines are gaining ground upon us, when we reflect on the increasing number of those who are bound to them as slaves, and of those who devote their whole souls to the advancement of the mechanical kingdom?

A class of objects in the nonhuman world that is somewhat analogous to cultural artifacts is the viruses. These are tiny strands of nucleic acids, often covered with a thin layer of protein, that reproduce only inside other living cells, either the prokaryote cell of the bacteria or the eukaryote cell of organisms like ourselves. Some are without dramatic effect, others cause such horrifying diseases as smallpox, polio, and yellow fever. I said they reproduced inside a living cell. This is not quite accurate. What happens is that a single virus particle enters, and then forces the cell to turn its entire metabolic apparatus toward the manufacture of the nucleic acids and proteins of a swarm of new virus particles. These often kill the host cell and burst out of the dead cell wall. Each survives, perfectly inert, not exhibiting the slightest evidence of life, until it encounters a new cell.

Like a cultural artifact, the virus causes its environment to produce more of itself. In some instances where the virus is not lethal, it alters its host in such a way as to benefit the virus. The bacterium of diphtheria does not cause disease unless it is itself infested with a virus. The virus alters the host metabolism so that it produces a toxin. The bacterium multiplies mightily in the diphtheria victim, all the while turning out a moderate number of viruses, not great enough to kill the bacterial cells. By analogy, the artifact may be said to control to some extent the behavior of its host, the social organism that is mankind.

If the community of scientists was somehow given the means to make a single expedition back into time, and had to restrict it to samples of a rather narrow interval, say 100,000 years, what time should they choose? Probably the best choice would be late Pleistocene, with several landing sites between 200,000 and 300,000 years ago. The team should be picked so as to best understand the ecology, biology, and psychology of that rare primate, *Homo*. At that time, according to our best judgment, the selective pressures modeling the hereditary equipment of our species would be at their height, for *Homo erectus* would be evolving into *Homo sapiens*. The definitively human component of our psychology that is instinctive would be (so far as our understanding of instinctive behavior now goes) genetically imprinted.

In the modern human population, the cranial capacity (brain size) ranges from 65 to 100 cubic inches (1,100 to 1,600 cc) in adults. The average is generally given as 84 cubic inches (1,375 cc). A fair-sized sample (some 50 or 100) of Neanderthal and Cro-Magnon skulls give an average of about the same value; these skulls are between 30,000 and 40,000 years old. Three very old skulls of *Homo sapiens*, approximately 200,000 years in age, are measured at 81 cubic inches (1,325 cc). Twelve specimens of Peking Man, who is classified as *Homo erectus*, dated at 400,000 years B.P. had an average brain size of 61 cubic inches (1,000 cc). Eight somewhat older skulls of *Homo erectus*, clustering around 600,000 B.P. and representing the Java man or close relatives, measured 55 cubic inches (900 cc). It can be seen from these data, which include all the skulls that had been found by the early 1960s and were complete enough to yield measurements, that there is little solid evidence for the "two plateau" hypothesis discussed earlier. It is certain that there has been little change in the average brain size of a relatively small sample of human beings from 30,000 to 40,000 years ago and an enormous, world-wide sample of humanity taken today. The very small sample (only three specimens) from nearly a quarter of a million years ago is not entirely convincing, but seen in its larger context seems to show that selection for larger brain size has not been in operation during this long period of human history.

There are two main theories to account for the presumed lack of evolutionary improvement of the brain during the last few hundred thousand years. One is that the physiology and mechanics of the brain make it impossible for evolution to further increase size. In order to function properly, the brain has to be kept at a stable temperature that can fluctuate no more than 3–4° F (1–2° C). Quite possibly the design difficulties are such that a larger mass of blood vessels and nerve tissues cannot meet this requirement.

The second theory is that cultural evolution has superseded and outmoded biological evolution, as it relates to the brain. One version is that in a society rich in material wealth, with good control over the environment, there is no continuous, widespread selection pressure for increased brain size. Another possibility is that the human species is so varied, has occupied so many different environments, and engages in such a variety of enterprises that there is no single target for natural selection. Were human populations to remain fixed, and were crafts and skills confined to genetically isolated populations over long periods, there might be selection for different kinds of brains, but it seems that this has not happened. Such pressure would not necessarily be for increased size of the cerebrum, the part responsible for the brain's large size in human beings.

Or it may simply be that the brain has already become so large as to produce the greatest degree of madness consistent with the biological survival of the human species under modern conditions. Put somewhat more kindly, the human mind is a frail instrument.

Some evolutionary changes are occurring in the human population before our eyes; ironically, they are not well understood. One is the increase in average stature, worldwide. Another is the younger age at which females reach puberty. It has been suggested that the greater mobility of the population and resulting marriages outside one's own local group have stirred up the gene pool at an increased rate. This may produce the phenomenon known to plant breeders as hybrid vigor, where hybrids are larger and faster growing, if given optimal conditions. Reasons for earlier sexual maturity are unclear. Some of these changes may be partly cultural (nutrition, or some other economic-related cause) rather than genetic.

A fairly well documented structural change in the transition between *Homo sapiens* and *Homo erectus* concerns the teeth. Lower jaws of *erectus* have a healthy, solid look. The third molars are heavy and well set. In modern skulls, the lower jaw has a weak, degenerate appearance. The teeth, especially the molars, are smaller, more variable, and more poorly set. Apparently the degeneration of the third molar began after the origin of *sapiens* and probably is still going on today. The obvious theory is that this results from weakening selection pressure favoring strong grinders, as food gets cooked and therefore softer. In the process the lower jaw gets shorter, so that there is less room for the third molar, or wisdom tooth, which is often either defective or missing.

Dr. M. Hellman, a dental surgeon at Columbia University's medical school and an associate of the American Museum of Natural History in New York, studied (1936) the dentition of 735 skulls in the collections of two large museums. These specimens represented populations from several localities throughout the world. Of the entire sample, nearly 28 percent of all adult males and 35 percent of all adult females had one or more wisdom teeth missing (never developed). Another 9 percent of males and 24 percent of females had one or more wisdom teeth impacted (abnormally developed so that they did not erupt). Hellman's work found no skulls with missing third molars among aborigines of Tasmania; among West African blacks between 2 and 6 percent had missing third molars; and 49 percent of European whites in Hungary lacked at least one wisdom tooth.

The evolutionary approach to an understanding of human beings had its effective scientific beginnings with the success of Darwin's theory of evolution by natural selection. The idea of a historical background to human behavior was particularly appealing to Sigmund Freud (1856–1939). His theories imply a layered sequence of mental activity, conscious and subconscious, which is both phylogenetic (evolutionary or historical) and ontogenetic (arising during development of the individual).

Freud was not a good biologist, however, and naïvely assumed that learned behavior or traumatic experiences could be inherited genetically. But, to his credit, he was dealing with mysteries which only now are being probed by advanced techniques of psychobiology.

Anthropologists are trying to recreate the precivilized era which occupied 95 percent of human history. They are doing this as archeologists, using shovels and trowels. Human bones and teeth, those of associated animals, pots, weapons, and other clues as to the mode of life of preliterate peoples are studied with increasingly sophisticated techniques. And, pressed for time, the anthropologists study the living primitive peoples, the small people of the Kalahari, the forest pygmies, the Australian bushmen, or the jungle dwellers of the Amazon basin. These cultures are vanishing day by day. They are as fragile as mist. Individuals are killed by the virus of an ordinary cold brought in from the outside. Their culture may languish after a gift of a few metal tools, or even after the trauma of talking with an anthropologist and learning something of the outside world.

It may be that even precivilized humanity disrupted the environment more than would be expected from a rather scarce mammal. The prehuman species *Homo erectus* knew how to use fire. In his time, and that of early humans, grass and forest fires apparently increased manyfold over lightning-ignited conflagrations. The intelligence, sharp flint blades, and throwing sticks of *Homo sapiens* of some 15,000 to 10,000 years ago are believed by some authorities to have been mainly responsible for the extinction of the so-called megafauna of the Pleistocene. These were the mastodon and rhinos of the northern forests, the mammoths of savannahs and grasslands, the elephant-sized ground sloths and giant armadillos, and the saber-toothed tigers and other carnivores that lived on the giant herbivores. In all the lands except Africa, these oversized mammals, few in number but providing an extraordinarily rich source of food to these skillful and brave early hunters, dwindled away to nothing at about the time that *Homo sapiens* turned to big game hunting. In North America, after exterminating the mammoths and mastodons, the Indians turned to the herds of bison. Ten thousand and more years ago, Indians killed bison that were giants compared to the modern forms, some with a horn span of 6 feet (2 m). Ingenuity in managing huge beasts by learning their habits and organizing this knowledge in scientific fashion, ingenuity in managing materials to construct traps and make blades more deadly than the long scimitars of the saber-toothed cats: these characteristics were prophetic of the innate cunning of the human animal which was to make it the ruler of the earth in another second of geological time.

Geologists have different opinions about how to name the more recent intervals of geologic time. In a recent textbook (*Evolution of the Earth,* by Robert Dott and Roger Batten), the "Age of Mammals" or Cenozoic Era is classified in this way:

Periods	*Epochs*
Neogene	Holocene or Recent Pleistocene Pliocene Miocene
Paleogene	Oligocene Eocene Paleocene

Some absolute ages, provided by radioisotope studies, are 65 million years B.P. for the beginning of the Paleogene, 25 million for the beginning of the Neogene. The

Pleistocene began 3 million years ago. The beginning of the Recent is taken to be marked by the most recent drastic rise of sea level, which coincided with the melting of the great ice sheets of northern Eurasia and America. This event took place between 6,000 and 8,000 years ago.

It is somewhat illogical to mark off the Recent from the Pleistocene. The latter is known as the Ice Age, a time when nearly all the fresh water of Earth was locked up in ice sheets that now cover only Greenland and the Antarctic. Once the ice sheet extended across Canada and down as far as Columbus, Ohio. Despite the warming phase that began the Recent epoch, we are still in the Ice Age. The earth will not return to normal climatic conditions until the Greenland and Antarctic ice caps melt.

Homo sapiens emerged as a species during the latter part of the Pleistocene. The beginning of the Recent epoch is not far from the beginnings of large-scale agriculture and of civilization. Some writers feel that the Age of Mammals is being replaced by the Age of Man. Whether this most recent Age turns out to be only an Episode remains to be seen.

It was the small island kingdom of England that, in the middle of the eighteenth century, turned loose the human species as a force of truly geological magnitude. The Industrial Revolution produced an animal species that was sustained by coal and oil, carbon compounds that had accumulated in the rocks as fossilized sunlight for hundreds of millions of years.

The Industrial Revolution did not emerge from nothing. For 2,000 years slag heaps had been accumulating from Britain to Persia, refuse from smelting ores of silver, lead, tin, copper, and iron. The forests of the Mediterranean lands had provided the fuel.

One writer says of England that at the beginning of

the eighteenth century the intellect of thoughtful Englishmen had applied itself to speculative problems of religion and philosophy. But by the middle of the century it applied itself to practical problems of industry. Adam Smith's *Wealth of Nations* appeared in 1776. In the year 1700 about 2.5 million tons of coal were mined in the United Kingdom; in 1800, 10 million tons; in 1900, 225 million tons.

By the early 1800s the United States had joined the fossil fuel club, and after 1870, the German nation had become, along with England, a fearsome converter of iron and coal into big guns and warships.

Production of iron was easy for England, which had abundant deposits of high-grade coal and iron oxide. Isolation of the English from the religious wars on the Continent gave them both freedom for long-range development and a supply of technically trained refugees from those wars (mainly weavers and potters). The Industrial Revolution was concerned about equally with textiles and armaments. For thousands of years women had been twirling spindles and spinning wheels to provide artificial fur to cover their naked families. English inventors during the last half of the eighteenth century, using the expertise of the weavers of an earlier generation, solved the problem of spinning and weaving on a large scale using steam power, iron machinery, and unskilled labor. Textiles thus became large-scale heavy industry. The technologist's political economy became the militarist's overseas empire, both to get raw materials and to break up the more primitive textile economies to open markets for manufactured goods.

Iron and steel for huge naval rifles and ground artillery, armor plate that was yards thick for battleships, and, in later years, energy for the conversion of atmospheric nitrogen into high explosives, all depended on fossil fuels. The fires of the smelter and of the chemical works increased by several orders of magnitude the energy that civilization used.

Roaring furnaces of steel mills and power plants seem to dwarf the quiet, cool, and invisible biological fires that keep alive the plants and animals around us. Just how important are the fires of industry as compared with biological activity? The magnitude of the fire, that is, the amount of energy it produces, can be measured by the amount of carbon produced as carbon dioxide.

The amount of carbon returned to the atmosphere from the sea as a result of the metabolism of marine organisms has been estimated, in one study, as about 100 billion tons a year. The amount from organisms on the land is about 35 billion, bringing the total to 135 billion. Combustion of coal, oil, and natural gas in all the factories, houses, and automobile engines of the world produces only 5 billion tons of carbon dioxide. Thus, biological production of energy is about 25 times as great as the "cultural" energy based on fossil fuels.

When we look at the effect of modern technology on the environment in more detail, we get a different picture. About half the earth's land surface is "wasteland," without commercial value except for tourism or minerals. Of the remaining half about 20 percent is tilled, 30 percent is pasture, and 50 percent is forest. Thus, the natural vegetation of a tenth of the land surface has been completely destroyed by plowing. Pasture land is usually overgrazed, so that the original vegetation has been replaced. Often what remains is mostly thorny, repellent scrubby plants that can be eaten only by goats or camels. Forest land is attacked in various ways, either for wood or to convert it into tilled land. Today, the greatest of the remaining forests, that of the Amazon basin, is literally going up in

smoke at the rate of tens of thousands of acres a day, in
the hope that it can be converted into grassland for cattle,
and in the certainty among its developers that its commercial value has appreciated ten or a hundred times by being
converted from jungle fit only for growing vines and Indians into bare red dirt with a problematic future.

In the geologic past, whole floras and faunas have
been profoundly changed, with many groups of animals
and plants exterminated. This happens, for example,
when continents drift halfway across the world, through
different climates. Sometimes they collide with alien continents having vastly different biotas which then come
into competition, with a good deal of loss of less well-
adapted kinds. Yet through all these changes the land
generally has had a cover of plants and animals that remains variegated, with many species.

Cultural changes in modern times have been more
profound in that they occur rapidly and tend to eliminate
diversity and substitute simple systems for complex natural ones. In the real world, with its unmanaged and unpredictable climate and surrounding biological hazards,
the simple systems tend to be difficult to manage. There
are in nature a myriad of factors which interact to correct
imbalances. The combination of a few species of crop
plants and heavy applications of insecticides to kill the few
species of pest insects produces a simple system that is
highly unstable. Climatic changes may destroy the crop or
enhance the growth of the pest. Use of a variety of crops
or insect enemies of the pest species would tend to com-
pensate for the unpredictable fluctuations of weather.

It may be that, for continued success, the agriculture
of high technologies will be forced into more complex and
sophisticated systems that use biological rather than sim-

ple chemical and mechanical processes. In *The Insects* (1964) I wrote:

> It often is said that we are now in the age of science, that this is an age in which science dominates all, and it is pointed out that most of the scientists who ever existed are alive today. But it may turn out that this is only a crude and trifling beginning of an age of science. The reason for thinking this is that current science is largely physical science, and the purely physical universe and the science that manages it is in a way extremely simple as compared with the biological universe and the biological science that can be developed in relation to it. An investment of human effort in biology commensurate with that of the physical sciences and with the inherent complexity of biological materials would probably require the cooperation of most of the human population. There is room not only for abstract mathematical thinking but also for the warm intuitiveness of the gardener, the hunter, the observer of living animals. And above all, since man is part of the biological universe, there would be need for a nearly incomprehensible degree of humaneness, in the face of the development of techniques that can change man himself.
>
> When one sees the wasteland produced by the extension of military and business ethics into the universities and research laboratories, one cannot contemplate with any pleasure the emergence of biology as a major activity if it is to be organized along similar lines. To have the spokesmen of haste, destruction, and profits take over the area that ties man to the living world that gave him birth would be an affront of a very special kind and significance.

Even now the worldwide effects of using fossil fuels are of interest to students of the planet earth. The problem becomes even more interesting if the projected doubling of population and rise of standard of living of low-technology countries actually occurs. According to one widely quoted set of estimates the carbon dioxide concentration of the atmosphere has between 1860 and the present increased from 290 parts per million (ppm) to 320 ppm, about 10 percent. By the turn of the twenty-first century, if only the combustion of fossil fuels is involved, it will be about 400 ppm. Carbon dioxide blocks the heat waves reflected toward outer space from the sun-warmed earth; some theoreticians have predicted that even this small accumulation of carbon dioxide will heat up the earth. However, the situation is complicated, and no consensus has developed on the effects of carbon dioxide accumulation in the atmosphere. Increased carbon dioxide production might increase plant growth, which would presumably lock up more carbon in the standing crop of vegetation, again removing it from the atmosphere. It is quite possible that in the fairly recent geological past there have been carbon dioxide fluctuations from natural causes that dwarf anything we have been able to produce by burning coal and oil.

To a large degree the iron or steel age has been replaced by the plastic age. Soybeans and petroleum are converted into long-chain molecules that are rather remote analogues of wood, fingernails, or silk. They range from soft, flexible materials to metallike substances that can replace steel. In their manufacture, organic molecules are produced as waste products, a few of which are extraordinarily dangerous to living things. One of their most insidious effects is to accelerate the subtle processes that convert a normal living cell into a cancerous one. Often the complete transformation into a cancer cell takes many years, so that the effect of the pollutant molecule is not immediately evident.

If everyone in the world used oil at the same rate per capita as the Americans, the world's oil supplies would be gone in a few years. Coal supplies are much larger, but are also finite. Planners who look into the future predict a world population stabilizing at 10 or 12 billion, about three times as large as that now existing. If this is accurate, then the fossil fuel stage of history will have to be only a brief episode.

Freeman J. Dyson wrote in the *Scientific American* (September 1971) that proof of a beneficent universe ("it almost seems as if the universe in some sense must have known that we were coming") lay in the fact that the human species will never run out of energy. When it outgrew supplies of wood, then coal and oil became available. Now that these fossil fuels are dwindling, we find that energy from atomic fission is available. Should the rare fissionable heavy atoms be used up, the fusion energy from hydrogen isotopes will last forever.

During the 1950s and 1960s many economic planners spoke confidently of the impending transition from the age of fossil fuels to the age of atomic energy. At a small and elegant museum built as a public education effort by the owners of a nuclear power plant was a bicycle, mounted on rollers, where the visitor could gauge his own energy-producing ability. Four electric light bulbs, rated at 25, 50, 75, and 100 watts could be lighted up by pedaling at the appropriate degree of effort. A young woman used to cycling managed, with extreme effort, to get the 50-watt bulb turned on. I, at 50 years and with no recent cycling experience, but driven to maximum effort by an amused audience, turned on the 75-watt light. Just a few pennies worth of electricity would have kept this light on for the whole day. The energy turned out by our gigantic power plants is essentially free if we are buying the amount that can be produced by one person.

However, the average use of energy per day for every child, woman, and man in the United States is 10,000 watt hours (10 kilowatt hours). Most people in the world use daily only the 1 to 2 kilowatt hours needed for biological metabolism plus perhaps a few more in a small wood fire used for cooking. If this majority of the population were to consume 10 kilowatt hours per day, the total energy use would be 40 billion kilowatt hours.

It is theoretically possible to find this much energy in the world supplies of fossil fuels for a few centuries. But in the real world, the scenario would probably go something like this. The growing scarcity, first of natural gas, then of oil, and finally coal would drive up the cost. This would drive down the standard of living of the poorer segment of the population. The net result would be merely to accelerate the growing divergence between rich and poor. About half the poor are already seriously undernourished. The percentage would continue to rise. Ill health or direct starvation would drive many to premature death, so that the projected world population of 10 or 12 billion would never be realized. A thin layer of corrupt privilege would ride uneasily on the confused mass of disintegrating humanity.

In place of this gloomy outlook, the sanguine engineer sees a transition from a fossil fuel economy to a nuclear energy economy that could support an entire world population of 10 billion with all citizens at a higher material standard of living than the present American average.

As is well known, the sun is a nuclear furnace, sending out a torrent of energy, warming our planet and turning the wheels of photochemistry in green plants that fabricate nutrients for the entire living world. By some perhaps inevitable coincidence, the atom is a miniature solar system, with a heavy central body, the nucleus, corresponding to the sun. Also, the atomic nucleus is the main energy source. In contrast to the sun, it is usually dark and quiet. But it conceals enormous stores of energy.

The ordinary energies of fires, automobile engines, or a runner depend only on rearrangements of the electrons orbiting the nucleus. The amounts of energy involved per atom are infinitesimal in comparison with the energy released by nuclear changes. The high temperature—18–36 million degrees Fahrenheit (10 to 20 million degrees Centigrade)—and crushing pressures in the interior of the sun bring about changes in atomic nuclei. That is, inside the sun nuclear rather than electronic chemistry is the norm.

On our quiet planet, nuclear changes are rather infrequent. They were not identified or understood until the early 1900s. There are a few naturally occuring elements that constantly turn out measurable amounts of heat and other energy in the form of "radiation," which consists of high-speed subatomic particles or short, high-frequency electromagnetic waves. Such elements were called "radioactive," and were shown to have unstable, changeable, nuclei. An exceedingly powerful natural radioactive element is radium, isolated by Marie and Pierre Curie at the end of the last century. An ounce of radium at a distance of 1 yard will give a lethal dose of radiation to a human being in about half an hour. (A modern nuclear power plant with a power of 300,000 kilowatts produces radioactivity equal to about 1,000 tons of radium; this radioactivity is kept within the fuel chamber by shielding.)

Early in the twentieth century it was discovered that a subatomic particle called the neutron could release atomic energy upon impact with a nucleus. This led by the 1930s to the regular use of neutrons in splitting uranium nuclei, with the release of both energy and more neutrons. The theoretically possible chain reaction first occurred on a reasonably large scale at the University of Chicago in 1942. Immediately the United States and

England began to produce a bomb utilizing such a reaction, and two (one of uranium, one of plutonium) were exploded over Japanese cities a few days before the end of World War II.

Production of such fission bombs (using the artificial element plutonium as well as uranium) went on at an accelerated rate, at first in the United States, then in a few years in the Soviet Union, and now in several countries, both known and unknown to the general public.

The explosion of a fission bomb produces sunlike temperatures and pressures, making it possible to mix the fissionable materials with such substances as lithium to produce super-bombs, the so-called hydrogen bombs, in which nuclei fuse to produce energy. Authoritative, detailed assessments of the physical effects of the explosion of all these bombs in various patterns within a short time have not been made public, except in general terms of expected casualties, which would be in the hundreds of millions. Despite the horrors, these would make only a small dent in the world population of thousands of millions. What is most significant is the long-term effect of widespread radiation, of which we know little.

Of general interest is the increase in background radiation that would be caused by all-out nuclear war combined with the release over a period of time of radioactivity from commercial plants left without supervision owing to the disorganization resulting from such a war. Such estimates do not seem to be available, but if projections for the next few decades on expansion of nuclear plants are realized, potential radiation levels could perhaps become incompatible with the existence of large vertebrate animals over most of the land surface. Insects and other smaller organisms are relatively resistant to radiation.

According to a report by the director for international

relations of the Rockefeller Foundation, about 215 million gallons of high-radiation-level atomic wastes have been produced by the nuclear reactors used in the United States to make plutonium for nuclear weapons. Of this a small fraction (430,000 gallons) has accidentally leaked out (through some 18 holes or cracks) of the temporary storage tanks. Another 600,000 gallons of commercial atomic wastes are being stored in temporary tanks. The authorities say a permanent method of storage does not have to be discovered until the early 1980s. Once the radioactive processes going on in these hot liquid wastes have been started, there is no way of stopping them. They must run their course, gradually decreasing in lethality over the centuries and millennia.

The history of cultural development of the human race has been one of trial and error, just as the evolutionary history of life has been one of the successful and unsuccessful lives of the myriad organisms that have lived and died in the brave and carefree way of wild creatures. In human affairs, this trial-and-error method was successful in post-Renaissance Europe where a large number of cultural groups, with an ideal combination of communication and isolation, worked out a variety of methods and attitudes in the development of technological societies. Mistakes in one region were not irreversible; the lessons from success in another country flowed into the vacuum, and what was once a backward region might take its turn as a leader.

The problem with the nuclear energy trial or experiment is that it may well not be reversible. New breeder reactors are required to counter the shortage of naturally occurring uranium 234, of which there is enough to fuel ordinary reactors for only a few decades. Breeder reactors convert thorium and nonfissionable uranium, both relatively abundant metals, into fissionable reactor fuel. A program for building large numbers of these breeders or fast reactors will produce plutonium by the hundreds of tons a year. The plutonium itself is exceedingly dangerous because of its initial reactivity and because of its tendency to catch fire if it is ground or disintegrates into fine particles, allowing radioactive debris to spread in uncontrollable fashion. The unalterable sequence of its breakdown into a number of radioactive elements produces a mixture that does not become reasonably safe for some quarter of a million years.

Already there probably exist tens of thousands of pounds of plutonium (a little more than 11 pounds, or 5 kg, is the critical mass for an explosion) loose in the world in irregular political or commercial channels. It is not at all unlikely that the multiplication of chaos in the management of nuclear materials will proceed as inexorably as the radioactivity of the materials themselves.

One of the most dangerous aspects of the development of nuclear energy is the decision of the military throughout the world to arm on a massive scale with small tactical nuclear weapons (those useful on a battlefield or in a naval engagement). These lie nearer the threshold for actual use than do strategic weapons. Vietnam showed that a determined population armed with small automatic weapons and hand-held guided antitank and antiaircraft weapons could not be defeated by conventional weapons. Hence the potential appeal of a shift to nuclear weapons; but there is no guarantee against an enormous proliferation of these small tactical nuclear weapons and their use in guerrilla warfare.

The most telling argument against major reliance on nuclear energy for power is the economic failure so far of energy plants, with heavy government subsidies required and a worsening economic future in nearly every forecast. The nuclear plant near the 75-watt bicycle mentioned earlier came into operation behind schedule, the contractor for supplying new fuel has reneged, and the plant is said to represent the greatest cost overrun of any major engineering project in the United States. However, experience has shown us time and again that profit can and will be made, in the short term, on economically improbable or impossible projects. Billions of dollars worth of unused and unusable aircraft are lined up in neat rows, shining under the sun of the American deserts. The jungles of the Amazon burn for immediate profit, plunging us into the consequences, which are almost certainly irreversible, of our dark ignorance of the ecological effects.

The basic cause of the present military and energy crisis is the immense increase in the human population. The immediate cause of the recent abrupt upturn in population density, in the underdeveloped nations at least, was the invention of antibiotics in the early 1940s. A doctor from the University of Michigan, who spent much of his professional life in Puerto Rico as a public health official, said upon his retirement that he feared he bore much of the responsibility for changing Puerto Rico from a paradise into a living hell—he considered his work to have helped produce the catastrophic rise in the population of the island.

The roots of overpopulation were well established much earlier. One of our greatest areas of ignorance about the mode of life of ancient human populations is the age structure. We infer a high infant mortality from our studies of, say, eighteenth-century Europe or of a semicivilized African or early American people who have come in contact with diseases generated by high-density populations elsewhere.

It is quite possible that early peoples, like most large mammals, had few young. The rabbitlike breeding of the inhabitants of early industrial England, for example, is abnormal, and probably only compensated for the high infant mortality caused by crowding as well as general demoralization of the working population.

Alexander Carr-Saunders, in his *World Population: Past Growth and Present Trends* (1936) writes:

> There is abundant evidence that there existed among primitive races what might fairly be called a small family system; that is to say certain customs were extensively and regularly practiced which, whether by limiting conceptions through restrictions on intercourse or by destroying the products of conception by abortion or infanticide, kept the number of living children small. This system, if it may be so called, broke down in Europe and the Near East with the rise of the early civilizations.

Women of the so-called primitive societies have a very sophisticated appreciation of the carrying capacity of their environment and of the means of controlling their production of children. The primacy of women in determining optimum birth rate was one of the first casualties of civilization. In the West, this disability was brought about largely through the activity of organized religion. Only now are liberated women beginning to reassume the responsibility which is basic to the survival of the human race.

In an overall view of the history of warring civilizations, men can be regarded as playing primarily the role of expendable mental defectives. Strangely, these defects proved to represent a preadaptation to the requirements of empire and industrialism. Physics, which developed in response to the need for weaponry, is a simple science compared to biology or social interaction, and the elevation of physics to the status of the model science is a serious error.

If the emergence of women continues, and men to a larger degree can (like the artist) keep some of their childhood openness and can learn from women, perhaps the wholeness of humankind can be restored.

Is *Homo sapiens* an endangered species, as some pessimists think? Those much-publicized birds and mammals that we know to be on the edge of oblivion usually exist in populations of a few hundreds, or even less than a hundred. The immense number of human beings would seem to make them invulnerable. Yet experience shows us that huge populations may suddenly become extinct before our eyes, without our understanding exactly what happened. Such were the passenger pigeon and the migratory locust of the American West. The initial decline of both was clearly and directly caused by humans, but both would be expected to continue to exist in small numbers in sheltered areas, allowing the species to continue in a more modest fashion. However, it seems that the biology of both species was predicated on life in a massive population, and that solitary or near-solitary individuals lacked the behavioral patterns to perpetuate life. So variable is the human species that it is difficult to imagine it driven into such a corner. It would seem that human history will stretch on into an indefinitely long future, through times both difficult and dangerous, safe and bright.

Earth, even if crippled with poisons of a failed episode of human history, will continue to turn under a beneficent sun and nourish the elegant tracery of life on its surface.

Index